Canto is an imprint offering a range of
titles, classic and more recent, across a
broad spectrum of subject areas and
interests. History, literature, biography,
archaeology, politics, religion, psychology,
philosophy and science are all represented
in Canto's specially selected list of titles,
which now offers some of the best and
most accessible of Cambridge publishing to
a wider readership

Beyond Science

Science is very successful in discovering the structure and history of the physical world. However, there is more to be told of the encounter with reality, including the nature of scientific inquiry itself, than can be gained from impersonal experience and experimental test alone. *Beyond Science* considers the human context in which science operates and pursues that wider understanding which we all seek. It looks to issues of meaning and value, intrinsic to scientific practice but excluded from science's consideration by its own self-denying ordinance. The author raises the question of the significance of the deep mathematical intelligibility of the physical world and its anthropically fruitful history. He considers how we may find responsible ways to use the power that science places in human hands. Science is portrayed as an activity of individuals, pursued within a convivial and truth-seeking community.

This book neither overvalues science (as if it were the only worthwhile source of knowledge) nor devalues it (as if it were to be treated with suspicion or not taken seriously). Rather *Beyond Science* provides a considered and balanced account that firmly asserts science's place in human culture, maintained in mutually illuminating relationships with other aspects of that culture.

John Polkinghorne has had a distinguished career as a particle physicist and as an author of books exploring themes in science and religion. He was elected a Fellow of the Royal Society in 1974 and was Professor of Mathematical Physics at the University of Cambridge in 1968–79. In 1982 he was ordained as a priest in the Anglican Church. Since 1989, John Polkinghorne has been the President of Queens' College, Cambridge. His books include *The Particle Play* (W. H. Freeman, 1979), *The Quantum World* (Longman, 1984), and *Science and Christian Belief* (SPCK, 1994).

To our children
Peter, Isobel and Michael;
and
our children-in-law
Jill, Timothy and Karen

JOHN POLKINGHORNE

Beyond Science

Science

The wider human context

CAMBRIDGE
UNIVERSITY PRESS

Published by the Press Syndicate of the University of Cambridge
The Pitt Building, Trumpington Street, Cambridge CB2 1RP
40 West 20th Street, New York, NY 10011–4211, USA
10 Stamford Road, Oakleigh, Melbourne 3166, Australia

First published 1996
Canto edition 1998

A catalogue record for this book is available from the British Library

Library of Congress cataloguing in publication data
Polkinghorne, J. C., 1930–
Beyond science / John Polkinghorne
p.   cm.
Includes index.
Romanized record.
ISBN 0 521 62508 4
1. Science – Philosophy. 2. Science – Social aspects. I. Title.
Q175.P834 1996
500–DC20   96-13177 CIP

ISBN 0 521 62508 4 paperback

Transferred to digital printing 2002

WV

Contents

Preface

I have had a long career as a physical scientist and I want to attend with great seriousness to what science can tell us. I have had an even longer career as a human being and I want to integrate my scientific insight into the broader and richer setting of personal life. This book is intended to look beyond science to the human context. It is concerned with individuality and community, meaning and value, and the springs of responsible action. In this endeavour I seek to tread a path between an exaggerated view of science's importance, as if it were the only source of reliable knowledge, and an impoverished view of science's importance, as if it were to be devalued lest it should rise above itself by turning into scientism, the proclamation that science is all.

Much of the material of the book is drawn from the New College Lectures I gave in Sydney at the University of New South Wales, together with some additional matter derived from the Sproule Lectures that I gave in Montreal at McGill University. I have subtracted from the latter source some of the theological reflections that were appropriate to its setting in a Department of Religious Studies. The substance of the latter thoughts can be found elsewhere in some of my other writings specially concerned with the interaction between science and theology. Nevertheless, I have permitted myself from time to time to make some reference in this present book to my religious beliefs and understandings where this has seemed directly relevant to its main themes. I am very grateful to the authorities at New College and at McGill for their kind invitations to give the lectures and for their generous hospitality on the occasions of my doing so.

My secretary, Mrs Josephine Brown, has once again provided all the skilled secretarial support that I needed, and I thank her warmly for her willing helpfulness. My wife, Ruth, has given valuable assistance in correcting the proofs, for which I am grateful, as I am also for the assistance of the staff of the Cambridge University Press in the preparation of the manuscript for publication.

John Polkinghorne

1

Is science enough?

A little while ago I was idly watching Australian television when a long interview by satellite with the geneticist Richard Dawkins appeared on the screen. The first half hour or so of the programme was devoted to a fluent exposition of Professor Dawkins' well-known bleak views of the meaningless character of terrestrial and cosmic history, which he sees as a concatenation of inane events whose only connecting thread is the propagation and survival of limited structure-carrying systems such as genes. In the closing minutes of the interview, however, the character of the conversation underwent a startling change. Dawkins explained that, though he might have sounded like an austere and somewhat desiccated character, he was in fact a warm person in whose life the experiences of human affection and scientific wonder played vital roles. He also called on us to transcend the narrow motivations of the selfish genes and to repudiate those notions of eugenics or racism that might have seemed to follow from a policy of genetic survival at all costs.

I do not doubt for a moment the authenticity of these closing remarks. A fully alive and responsible human being could hardly say less. Yet I longed for the interviewer to inquire where these humane and moral encounters found their lodging in the empty world, devoid of meaning, which the speaker had been describing. Richard Dawkins is celebrated as being the apostle of an extreme and unrelenting material reductionism.[1] In theory, for him science must be enough, because there is no other reliable form of knowledge to be had. The richness of experience and insight that makes human life worth living is, in his official view, reduced to epiphenomenal triviality.

We are entitled to require a consistency between what people write in their studies and the way in which they live their lives. I submit that no-one lives as if science were enough. Our account of the world must be rich enough – have a thick enough texture and a sufficiently generous rationality – to contain the total spectrum of human meeting with reality. The procrustean oversimplification of a fundamentalist reductionism will not begin to suffice. In fact, it cannot even embrace the practice of science itself, which calls for judgements of value (we seek elegant and economic theories) and whose chief reward is the experience of wonder at the rational beauty of the physical world. The discoveries of science place in our hands an enhanced power to do good or ill, and scientists have to participate in the human search for insight in order to act wisely and to avoid harm. The deliverances of science constrain our metaphysical understandings but they do not determine them. There is much else that must also be taken into account. The context of science is the human context; it is an activity of persons, involving unspecifiable powers of creative imagination. Science by itself is not enough even to describe the pursuit of science itself.

It is the purpose of this book to go beyond science in order to consider the wider setting in which its activity takes place. I greatly value scientific discovery, and the marvellous insights that it provides, but I could not possibly think science was sufficient on its own to quench that thirst for understanding which is so natural a desire of the scientist. The chapters that follow explore some relevant themes. The first must be devoted to a defence of science as a reliable source of knowledge within its own self-limited domain. Such a defence is certainly necessary in these post-modern times. Scientists, as scientists, may not know enough, but they have caught sight of one needful thing: there is a truth to be found. We should acknowledge that their discipline is a part of that truth.

NOTE

1.   R. Dawkins, *The Selfish Gene* (Oxford University Press, 1976); *The Blind Watchmaker* (Longman, 1986); *River out of Eden* (Weidenfeld and Nicolson, 1995).

2

Understanding the physical world

Looking for answers

One of the most impressive aspects of science is its power to provide universally satisfactory answers to the questions that it asks. When I was a young research student, many years ago, we supposed the fundamental constituents of nuclear matter to be protons and neutrons. In the course of the following twenty-five years a complex interplay of experimental discovery and theoretical insight led eventually to the identification of a new level in the structure of matter: the celebrated quarks, and the particles which make them stick together, which (I regret) are called gluons. There was much argument and perplexity along the way, but in the end the dust settled and we could all agree that a great new discovery had been made. This unanimity of conclusion is very convincing. It leads people to see science as real knowledge – indeed, perhaps, the only form of real knowledge, since such unanimity does not seem to be forthcoming in other domains of human inquiry, such as politics, ethics or religion. This leads to what one might call the hagiographical account of science: it is the canonical source of what we can know about the nature of reality. People who wish their belief to be based in the solid ground of fact should look to science alone and eschew the treacherous swamps of mere opinion that surround it. This strategy may very well lead to a diminished account of the world – for it deliberately sets aside almost all that makes human life worth living – but, if we are sternly honest, it is the meagre best that is to be had. That is how the hagiographical account sees the situation. Science is our only reliable guide in the perplexity of human experience.

While this way of evaluating science appears to be quite widely assumed, and while it certainly has some vigorous proponents among reductionist scientists (particularly biologists[1]), it is by no means the only evaluation on offer. Opposed to it is what we might call the deconstructive account of science. This comes in two forms, a popularistic (indeed often, journalistic) condemnation of the failure of science to meet human needs, and a philosophical critique of science's ability actually to deliver knowledge at all.

The first form sees the scientific imperialism implicit in the hagiographical account as being destructive of human good, with its depersonalization of reality and its rejection of value. Science furnishes humanity with great resources of power, but since it does not trade in wisdom, we are left with the actual ecological disasters, and the potential military disasters, for which technology has provided the catalyst.

What is at issue in this part of the critique is the question of scientism, not of science itself. The hagiographies accord so unique a status to science that its legitimate claims to afford understanding are blown up into a universal competence that it cannot sustain. Physics is illegitimately promoted into metaphysics. Scientism – the claim that science is all – provides a grotesquely impoverished account of reality, which its critics are right to reject. But in doing so, many of them seem to be in danger of rejecting also science itself.[2] Within its self-limited domain of inquiry (roughly, the world treated impersonally; reality in the objective mode of an 'it') science has much to tell us that we should take with the utmost seriousness. Or, at least, that is what I think, but the philosophical critics of science would, in their more sophisticated deconstructive way, beg to differ.

Philosophical debate

The twentieth century has seen a very vigorous debate in the philosophy of science.[3] Few agreed conclusions have emerged but I think that all participants, at least, would acknowledge that science, both in its method and in its achievement, is a good deal more subtle than might appear on the surface. In particular, the story of clear theoretical prediction receiving unchallengeable experimental confirmation and so leading to certain truth is altogether too simplistic a tale. We cannot avoid facing the many necessary considerations that seem to complicate the issue.

The first is the unsustainability of a clear separation between theory and experiment, so that the one cannot in fact unambiguously confront the other. In 1984, two experimentalists in my old subject of elementary particle physics, Carlo Rubbia and Simon Van Der Meer, received the Nobel Prize for their discovery of W and Z particles, the mediators of the weak nuclear force. It was certainly a splendid achievement, but how was it done? Their very large research team employed a vast array of electronic detectors, linked together so that their signals could be evaluated through computerized analysis. The raw material of the experimental result was the registrations in these detectors, but data in that form are of no immediate significance. It is only when the patterns are interpreted *using current theoretical ideas* that one receives the message 'Here is a Z' or 'There is a W'. In other words, theory and experiment are inextricably intertwined in the scientific acquisition of *interpreted* experience. All scientific observation is a form of 'seeing as'; we have to interrogate the physical world from a chosen point of view. Of course, that point of view may prove in need of correction, but scientists find such revisions as difficult and uncongenial as anyone else. In the middle 1950s great effort was exercised by many physicists in trying to understand some puzzling coincidences in the behaviour of meson decays. They believed there must be two sorts of particle involved because they seemed to be seeing two different kinds of behaviour under spatial reflection. Yet, all the other properties of these supposedly different mesons were exactly the same. After about two years of increasingly desperate but unconvincing attempts at ingenious explanation of this strange coincidence, two American-Chinese, T. D. Lee and C. N. Yang, made the brilliantly simple suggestion that maybe particles in these types of decay did not have to have a unique form of behaviour under reflection. That meant there could be just one kind of meson involved, after all. Lee and Yang had made the great discovery –

again of Nobel calibre – that what we call parity is not conserved in weak decays. They made their discovery by looking at those meson decays in a new way. They saw the physical world differently from the way in which their colleagues did. The need to adopt a point of view introduces a certain precariousness into the scientific enterprise. We have to beware of the tricks of an imposed perspective whilst being able to profit from the insight afforded by a well-chosen perspective.

Another problem arises from the fact that, though theory is indispensable to science, it is always underdetermined by data. After all, theories purport to speak in universal terms of what will happen at all times and in all places, but in our experimental sampling of physical experience we are only able to make limited encounter with what is going on. A particular aspect of this difficulty is the problem of induction, so trenchantly stated by David Hume in the eighteenth century. Why should past experience be the guide to future behaviour? The sun may have risen today, but I cannot *logically* deduce that it will rise again tomorrow. How many instances of Z-particle patterns did Rubbia and Van Der Meer have to see before they could really know there were such particles? In its assertion of universal knowledge, is not science going far beyond what it can, in all sobriety, actually claim to know?

A further problem is provided by the cloudiness of our vision, the impurity of our encounters with the physical world. Rubbia and Van Der Meer were looking for Ws and Zs, but the production of these particles was only a small fraction of what was going on in their apparatus. Other, much more copious, interactions may occasionally produce an accidental mock-up of an event that looks like, but is not, a W pattern or a Z. Then there are cosmic rays wandering in from outer space, impinging on the detectors and 'doing their own thing', which again may occasionally produce false signals. If physicists are to understand what is actually happening, they must be able to strain out these unwanted contaminations. In the trade, this is called the problem of eliminating background events. There is no rule book to tell you how to do it. Judgement has to be exercised in order to guess where trouble might lie and then theory called upon to estimate the consequences and necessary corrections. There have been a number of embarrassing episodes in physics in which totally incorrect conclusions have been reached through errors about background effects.

Once one paints a realistic portrait of science 'warts and all', it becomes increasingly difficult to accord it a unique status as the sole reliable source of human knowledge. Yet, one also recognizes the cumulative fruitfulness of the history of science whereby, for instance, the successive levels of the structure of matter – from atoms to nuclei to protons and neutrons to quarks and gluons – have been unravelled. Does it not seem difficult to deny that here science is telling us something to be taken absolutely seriously about the structure of the physical world in which we live? Not everyone, however, has seen it that way. Some reject what they term a 'Whig view of history' in the account of scientific progress, emphasizing instead the occasions of discontinuity, as when Ptolemy gives way to Copernicus, or Newton is replaced by Einstein and Bohr. For Sir Isaac, the physical world was clear and determinate, its processes taking place within the even flow of an absolute time. For his twentieth century successors, the physical world is cloudy and fitful at its quantum roots, and the elapse of time and the judgement of simultaneity depend upon the state of motion of the observer. Such radical revision seems to put in question a triumphalist claim for the steady advance of scientific knowledge. And who knows what further revolutions of understanding may await us in the future?

Science's achievement

This sketch of the history and practice of science disabuses us of the notion that science has a straightforward technique for the ascertainment of certain truth about the physical world. Any realistic account will have to be much more nuanced than that in its conclusions. Two great questions face us: How can we rightly assess the achievements of science? Is there some discernible essence of the way it goes about attaining that achievement? In other words, what does it discover and how does it make these discoveries.

It is clearly not possible to claim that science establishes truth, pure and simple. Entry into some hitherto unexplored regime (of higher energy or shorter distance or whatever) is always liable to reveal new and surprising phenomena that will call for conceptual modification or even a totally new way of thinking. In that sense, the conclusions of science are necessarily provisional. Indeed, it is the possibility of finding the unexpected around the next experimental corner that motivates the expensive exploration of artificially created regimes (such as are produced in high energy accelerators), far beyond what is accessible to us through natural encounter with the physical world. Yet, when we consider a regime that has been well winnowed by the flail of experiment and the sieve of theory, then we do not expect to have to change our ideas radically about what is going on. Newton did not say the last word about the solar system – Einstein's theory of a general relativity is necessary to explain the fine detail of the behaviour of the planet Mercury – but his theory of gravity is sufficiently close to what is actually the case for us to be able to use it to send a space probe to Mars. If science does not attain absolute truth, surely it can lay claim to verisimilitude. Its established theories give reliable accounts of what is going on in a carefully delimited domain, to specified degrees of detail and accuracy. Scientists are mapmakers of the physical world. No map tells us all that could be said about a particular terrain, but it can faithfully represent the structure present on a certain scale. In the sense of an increasing verisimilitude, of ever better approximations to the truth of the matter, science affords us a tightening grasp of physical reality.

So say I, and almost all other scientists with me, but not so all philosophers. The intertwining of interpretation with experience, together with the underdetermination of theory by experiment, has persuaded many of them that science's encounter with the physical world has about it a degree of elasticity that yields considerable room for explanatory manoeuvre. The theoretical insights of science are then seen as the imposition of a pattern of meaning on a veiled and elusive reality, rather than as reliable inferences from encounter with its actual nature.

The most extreme critics of this kind are those who see science as being largely, or even totally, a social construction. Thus, the sociologist Barry Barnes can propose that 'all knowledge generation and cultural growth should be regarded as endlessly dynamic and susceptible to alteration just as is human activity itself, with any actual change or advance a matter of agreement and not necessity'.[4] In his view, in the 1970s we did not *discover* quarks, we simply (unconsciously) agreed to view the world of ambiguous experience in a quark-like way. Physicists have the choice of what experiments are worth doing and how they should interpret them. In consequence, they can mould their encounter with the subatomic world into a shape that pleases their intellectual fancy. Anyone who does not go along with that self-imposed orthodoxy is excluded from the invisible college of scientists. That is the way the so-called 'Strong Programme' of the social determination of science sees the matter.

It is difficult to exaggerate how implausible this account seems to a high energy physicist. Far from the physical world proving to be like clay in our theoretical hands, it displays a diamond-like hardness, resistant to our expectations and imposing upon our minds its idiosyncratic and unanticipated structure. It is an immense struggle to find a theory that is economic and uncontrived and adequate to a wide swathe of experimental investigation. More than twenty years elapsed between Murray Gell-Mann's discovery of the strangeness quantum number and the articulation of the fully developed 'Standard Model' of the quark theory of matter.[5] They were years of continual experimental surprises and ceaseless theoretical struggles to make sense of what was going on. When finally a coherent picture emerged, it had all the feel of discovery and none of the feeling of pleasing construction. 'So *that's* what nature's like – who'd have thought it beforehand!'

Of course, it might seem possible that the physicists were mistaken and the philosophers and sociologists knew best, but I fear that the second-order commentators have paid too little attention to the accounts of their actual experience given by the first-order players. A part of that experience is the occasional critical experiment in which a definite indication from nature clearly points to the attainment of understanding through a particular kind of idea.

On the way to the discovery of quarks and gluons, there were crucial moments of insight derived from such critical experiments. One such was the discovery in the late 1960s of what is called deep inelastic scattering. When very high energy electrons are scattered by protons, some of them 'bounce back' in a surprising way. The historically minded would have thought of Rutherford and his colleagues in Manchester in 1911. Working at much lower energies, they had detected a similar bouncing back of α-particles when they impinged on a thin gold foil. Rutherford said it was as astonishing as if a 15 inch naval shell had recoiled on impact with a sheet of tissue paper. He went on to interpret it as indicating the presence of concentrated positive charge inside the gold atom. In a word, he had discovered the nucleus. It was possible to understand the Stanford electron experiments in a similar fashion, but now the little hard scattering centres proved to be identifiable with the quarks sitting inside the proton. Up to that point it had been possible to think of quarks as being no more than a kind of theoretical toy, a notional device to generate certain patterns in the ordering of matter; thereafter it became clearer and clearer that quarks must be reckoned as a new constituent level in the structure of the physical world. Some people liked that, others did not, but the physicists had been given a nudge by nature that no one could ignore, whatever their predilections. Of course, its recognition depended on interpretation. Matter is not stamped 'made of quarks'. But the interpretation was both so natural, and so effective in explaining the phenomena, that it could not be gainsaid. The origin of quark theory lies in the physical world and not in the minds of the physicists.

No one could deny, of course, that social factors operate in science. What experiments are considered worth doing (what experiments will in consequence be paid for), what theoretical ideas are fashionable (what in consequence most theorists will want to work on and solve) – all these are affected by social forces within the scientific community. I mentioned earlier the discovery of parity nonconservation. The experiments that confirmed Lee and Yang's ideas had been a practical possibility for many years. No one bothered to do them because they were considered uninteresting. Physicists thought they already knew what the answers would be. Such social factors certainly advance or retard the progress of scientific knowledge. But they do not determine what that knowledge shall be. When eventually the parity experiments were done, there could be no cavilling about the conclusion. Science is socially influenced but it is not socially constructed.

Paradigm shifts

More troubling, however, are those occasional revolutionary turning points in science that bring with them radical revision of our understanding of the physical world. When the nature of mass changes from being intrinsic and invariable (Newton) to being variable and motion dependent (Einstein), or the nature of causality changes from being deterministic (Newton again) to being probabilistic (Heisenberg), a much greater challenge is presented to claims of science's ability to give an account of an actual reality than is represented by the discovery that some previously supposed fundamental constituents are, in their turn, in fact composites. In the latter case, changing the scale of the scientific map has simply revealed the existence of more detailed structure; in the former case, the whole character of the terrain seems to have been altered – Newtonian *terra firma* has become a quantum quagmire.

Thomas Kuhn[6] has particularly emphasized this difficulty. He sees revolutionary turning points as involving a transition from one paradigm of the nature of the physical world to another. Kuhn exhibits a certain looseness in the use of his characteristic concept of a paradigm, but in essence he intends it to refer to a total account of scientific reality, with emphasis on the interpretative principles brought to bear in forming and defining that account. A paradigm is the way its proponents view the world and for Kuhn paradigms differ so greatly from each other that they are incommensurable. Like those ambiguous puzzle pictures beloved of gestalt psychologists (the duck/rabbit, the young girl/old crone) either you see it one way or you see it the other. There is no mediating position possible. In Kuhn's view there is a Newtonian world and an Einsteinian world, but they are so disjoint from each other that Isaac and Albert will be unable to hold a conversation in the Elysian fields.

If that were true, it would subvert not only claims of scientific verisimilitude but also claims of scientific rational motivation. If Newton and Einstein cannot talk to each other, there is no debate possible between them and their world views. We simply have to see which can shout louder. Kuhn made explicit comparison between the ways of scientific revolution through paradigm shift and the ways of political revolution. In his view, the issue was not truth but the effectiveness of ideological propaganda. (In his later writing he somewhat modified this extreme stance.)

Kuhn has proved widely influential outside science. The notion of paradigm shift has proved eminently exportable. Yet his account of science is not one that makes sense to a scientist. An essential feature of any scientific revolution is the successful construction of correspondence principles, providing the way in which the new theory can annex to itself the successes of the old, by showing the old to be the limit of the new in some well-defined physical regime. Newton was not abolished by Einstein but simply shown to be reliable only when velocities are small compared with that of light. The scale limitations of the Newtonian map were clearly established. Far from facing each other in mute and mutual incomprehension, Einstein and Newton are surely engaged in an animated discussion up there in Elysium. Sir Isaac will be interested to know that moving clocks run slow and so time is not as absolute as he had supposed. The ideas of that third class clerk in the Berne patent office did not prevail because he possessed a superior propaganda machine to that of the grand old men, Professor Lorentz and Professor Poincaré. Einstein's ideas prevailed because special relativity provided a more coherent and effective description of what was happening. The way cosmic ray particles decay shows that moving clocks really do run slow. Radical change is not revolutionary discontinuity. Interpretation is not a non-negotiable package deal. It is not the case that either you believe everything that Newton believed or you cannot speak to him at all. Certainly he and Einstein attributed different properties to mass, but both of them were speaking about *inertia*, the resistance of a body to having its state of motion changed. There is common ground on which both can stand.

Answering the critics

In rejecting these irrational accounts of the nature of science, we can also reject the impoverished assessments of the nature of scientific achievement to which they give rise. The minimalist account of positivism – that science is simply concerned with correlating sets of sense data – has long since been abandoned. It foundered on the failure to recognize that the intertwining of theory and experiment, fact and interpretation, denies us access to the unproblematic basic phenomena that the positivists assumed they had to hand. Even if that had not been the case, there would have been the further defect that positivism entirely fails to motivate adequately the scientific enterprise. It is one thing to expend much money and intellectual effort to find out what the world is like; it is something entirely different to do so simply to correlate experience, much of it of a highly contrived (experimental) character.

Today, critics of science tend rather to assert that its achievement is the instrumental success of getting things done. Pragmatism is the name of the game.[7] Scientific theories do not tell it like it is, but they are practically efficacious manners of speaking. Yet, whence arises this miraculous accomplishment if our theories do not reflect, at the least, some true aspects of the ways things are? The concept of electrons enables us to manufacture electron microscopes, understand superconductivity, explain chemical valency. These are rather remarkable achievements for a mere manner of speaking if it did not in some way resemble reality.

Technology may be about manipulative power, but science is about the search for understanding. I have suggested the following parable.[8] A black box is delivered to the Meteorological Office with the instruction 'Feed in today's weather at slot A and out of slot B will come the prediction of the weather in a fortnight's time'. Lo and behold, it works! The pragmatic task of the meteorologists is perfectly (if mysteriously) accomplished. Do you think they would all go home? Not a bit of it! They would take that box to pieces to find out how it modelled the great heat engine of the Earth's seas and atmosphere so accurately. As scientists they know that prediction, however perfect, is not enough. They want to *understand* the nature of weather systems.

Empirical adequacy – 'saving the phenomena', in the old phrase – is not sufficient. Bas van Fraassen tells us that this is all to which science can aspire. Its theories can only command acceptance, not belief.[9] I do not think we need to settle for that.

Of course, we cannot assert that science establishes truth, pure and simple. Its corrigibility, particularly in times of revolutionary change, means that verisimilitude – mapping within the limitations of a scale – is all that can legitimately be claimed. The defence of the proposition that this is what is happening seems to me basically to be itself an empirical question. We could not know *a priori* that humans would have this wonderful power deeply to understand the physical world, but as a fact of experience it has proved to be the case. The astonishing fruitfulness of basic scientific theories is one of the chief encouragements to seeing science in this verisimilitudinous way. Let me give you an example.

In 1928, Paul Dirac published what was probably the crowning achievement of his immensely fruitful career: the Dirac equation of the electron. He had been led to the discovery by the need to reconcile quantum theory with the special theory of relativity. The Dirac equation did this in a deep and satisfying way. It was a quite unexpected extra bonus, however, that it also explained a hitherto totally baffling aspect of the electron's behaviour, namely that its magnetic effects were twice as strong as one would have expected them to be. This property emerged in a perfectly natural, but unforeseeable, way from the equation. A few years later Dirac was led to the discovery of antimatter (that the electron has a 'twin', the positron, with which it annihilates to give a burst of radiation) by the need to find a way of understanding the negative energy solutions of his equation. Such remarkable fertility is surely a sign that one is in contact with physical reality. Our grasp of that reality is verisimilitudinous rather than absolute. There are further small magnetic effects that for their explanation require the more elaborate theory of quantum electrodynamics, to which the Dirac equation of a single electron is only an approximation.

It is experiences like this that encourage scientists to take a realistic view of their achievements and so to claim that we are learning what the physical world is actually like, There remains, however, the question of how we are to characterize the method of scientific inquiry that yields such satisfying knowledge.

Scientific method
Most scientists are philosophically unreflective but among the minority who are concerned with asking metaquestions about

the nature of their activity it is probable that Karl Popper[10] would be one of the first philosophers to whom they would refer.

Despairing of solving the problem of induction, Popper has emphasized that if we cannot know when a scientific theory is true we can at least know when it is untrue. His thought places great emphasis on the scientific role of falsification. 'All swans are white', until the first black swan found in Western Australia emphatically negates the proposition. Here, surely, is a very important component in any account of scientific method, namely its vulnerability to contradiction. Popper believes the essence of the procedures of science to lie in the pursuit of bold conjectures that are open to refutation in this decisive way.

This rings bells with many scientists, but on further reflection, I believe, one's enthusiasm for the Popperian point of view diminishes. For one thing, falsification is far from being an unproblematic concept. This is so even for low level observational theories like the whiteness of swans. After all, that black bird in Western Australia might be just a long-necked duck, and to rebut that suggestion one would have to move to a higher level discussion about the taxonomic classification of birds. For deep scientific theories, like quantum theory or special relativity, falsification is a very subtle matter indeed. Special relativity did not crumble when the respected experimentalist D. C. Miller reported the measurement of a non-zero aether drift. Einstein imperturbably said 'The Lord is subtle but he is not malicious'. There was too much going for his theory for it to disappear at the first hint of an adverse result, and he was right.

Also, science does not in fact progress by continually drawing a bow at a refutable venture. There is something much more skilful and directed about its exploration of the physical world. There is something very lop sided about Popper's account, with its emphasis on the certainty of falsehood but the inaccessibility of truth. In his view, Rubbia and Van Der Meer could never know for sure whether there are W or Z particles. What could have justified a party, in Popperian terms, would have been a null result, the falsification of the theorists' expectations. That would have been a real discovery.

There is something very odd about that conclusion. Popper is warmly appreciative of science, and when his heart takes over from his logical, deductive, head, he sees that this is so. There is in his writing an occasional hankering after the idea that repeated verification leads to ultimate confirmation – in plain words, that induction is not ruled out as a route to knowledge. There is clearly much more to be said about science than Popper has been able to articulate.

A considerable advance in relation to the question of the nature of falsification was made by Imre Lakatos[11], with his concept of a research programme. This provided a way of understanding how deep scientific theories can be sustained even if they do not always enjoy a perfect fit with observation. A research programme is defined by its central core of non-negotiable concepts, which defines the programme and which will be held on to whilst the programme remains active. For example, the hard core of the Newtonian research programme is the idea of gravity obeying a universal inverse-square law. Between the core and the phenomena is a protective belt of auxiliary hypotheses that are capable of adjustment in order simultaneously to preserve the core and to save the phenomena. These adjustments are not arbitrary but they are made in accordance with a strategy of explanation that Lakatos calls 'the positive heuristic'. For example, when the planet Uranus was found not to be fulfilling its Newtonian predictions, this did not lead to the conclusion that Newtonian gravity was false. Instead, John Couch Adams and Urban Leverrier introduced the new auxiliary hypothesis of an undiscovered planet beyond Uranus, per-turbing its motion through conventional gravitational effects. The subsequent discovery of Neptune was a 'novel, stunning, dramatic' success, which justified the programme being evaluated as being 'progressive'. Yet when discrepancies were detected in the be-haviour of Mercury, the Newtonian programme could not cope successfully with them by the auxiliary hypothesis of another undetected planet, Vulcan, this time lying close to the sun. There was no such planet. After 200 years of success, the programme had become degenerating. It was replaced by Einstein's General Relativity research programme, which not only gave a natural explanation of the behaviour of Mercury but also scored its own novel, stunning, dramatic success with the prediction of the deviation of starlight by the sun's gravitational field.

Clearly, there is a great improvement here, in that Lakatos's account gives a more recognizable description of scientific activity, but there must be more to be said. The Lakatosian pattern is too flexible; it can fit too many non-scientific examples for it to have distilled the complete essence of what is going on. Let me define the England Rugby XV research programme. Its hard core is the belief that they are the best side. I explain their defeats by my auxiliary hypotheses of unfortunate injuries, doubtful referee's decisions, and so on. From time to time they win the Triple Crown or beat the Wallabies or the All Blacks, and I score a stunning success. It seems clear that there are still some essential ingredients of scientific method which need to be identified.

The man who can help us here is, I believe, Michael Polanyi[12]. He has been strangely neglected by the philosophers of science, though he has many things to say that resonate with the actual experience of doing science. Perhaps neither of these facts is surprising. They stem from Polanyi's having himself been a distinguished physical chemist, so that he was an outsider in the philosophical community but an insider in the scientific community.

Polanyi's central thesis is that, though science is concerned with interrogating an impersonal physical world, it is itself an activity that can be pursued only by persons. Scientific knowledge is *personal* knowledge, because it is inescapably based upon acts of personal judgement and its pursuit requires a personal commitment to a point of view, even though scientific corrigibility means that that point of view could conceivably be false. In adjudicating questions of the adequate elimination of background effects, or the sufficient degree of attained verification of a theory, there is no rule book that the scientist can consult to settle the matter, no algorithm that would enable these questions to be delegated to analysis by computer. Yet scientific judgement is not a matter of idiosyncratic individual decision, for it is pursued within the convivial community of scientists and with universal intent. Similarly, formulating a new theory is an act of insight offered for evaluation by one's peers. It is not just read out of the data but it involves a creative leap of the mind. Polanyi's account strikes a judicious balance between the individual insight of the imaginative scientist and the essential receptive and critical role of the community within which the practice of science is necessarily pursued.[13] How that community functions is the subject of Chapter 3.

Scientists have learnt the tacit skills of doing science by a long apprenticeship within their truth-seeking community. They are content to submit their individual offerings to be sifted and assessed within that community, yet the paramountcy of the physical world as itself the ultimate source of knowledge preserves science from being merely a social construction.

The fact that there is no exhaustive specification of the essence of the scientific method is because of its irreducible character of personal knowledge. Tacit skills are real skills, whether they are those of riding a bicycle, or connoisseurship, or doing science. In an off-repeated phrase, Polanyi assures us that we all 'know more than we can tell'.

Critical realism

We have, in the course of the pursuit of truth, to be willing always to respect the nature of the object of our inquiry. Some may feel disappointment that scientific method cannot be given a crisper, more cut-and-dried, characterization, but I am persuaded that the chiaroscuro of the personal knowledge account is entirely in accord with the actual character of scientific activity. The view I am defending is called critical realism. 'Realism', because it claims that science actually does tell us about the physical world, even if it does not do so finally and exhaustively. 'Critical', because it recognizes the subtlety and ultimate unspecifiability of the scientific method.

If what I claim is true, two things follow. One is that science is not radically different from other forms of human rational inquiry. It too requires the act of intellectual daring, of commitment to a potentially corrigible point of view. It too involves reliable but unspecifiable acts of judgement. Science's superior power to settle questions lies, not in its invincible certainty, but in the openness to testing that results from its concern with aspects of reality sufficiently impersonal in their character to be open to repetitive investigation and consequent experimental checking.

The other consequence I wish to draw from the recognition of science as personal knowledge is that science's verisimilitudinous success encourages us to believe that rational, if precarious and unquantifiable, strategies of investigation of this kind are capable of leading us to an enhanced understanding of reality. One could not have deduced beforehand that this would be so (logicians such as Popper demand too much in that respect), but it is a contingent but fortunate fact about the world that we can learn much of its nature in this way.

Best explanation

In both experimental science (such as subatomic physics or biochemistry) or observational science (such as cosmology or animal behaviour), scientists are seeking the best explanation they can find of a great swathe of varied, and often puzzling, data. Such a 'best explanation' will be characterized by empirical adequacy, accord with general principles, economy, elegance, and long-lasting fruitfulness. The assessment of these optimal characteristics itself calls for acts of personal judgement, in which it seems to be the case that the scientific community can concur. It is an empirical fact of the history of science that time and again the scientists have been able to discover and agree upon just such a best explanation. Nature could have been opaque to us; it has not proved to be so. Given the underdetermination of theory by data, candidates for best explanation could have multiplied. This has again not proved to be the case.[14] Given the ingenuity and competitiveness of young scientists, always seeking to establish their reputation by the suggestion of some brilliant new idea, I cannot believe that this unanimity of conclusion on fundamental theoretical matters arises simply through slothful acquiescence in a socially agreed paradigm.

Others forms of human inquiry – and I would include theology on this list – are also seeking the best explanation of the experience they survey. They can take heart in their quest from the success of science, their intellectual cousin in the search for truth.

I write as someone who wants to take science absolutely seriously and to accord it its rightful place in the great human search for understanding. In my view, its achievement is a verisimilitudinous telling of what the physical world is actually like, in its structure and its history. Science's method is the pursuit of knowledge through acts of personal judgement within the conviviality of a truth-seeking community and in submission to the inflexibility of the way things are. Its relation to other forms of human inquiry is both comradely and encouraging. Science should be part of everyone's world view. Science should monopolize no one's world view.

NOTES

1. See, for example, R. Dawkins, *The Selfish Gene* (Oxford University Press, 1976); *The Blind Watchmaker* (Longman, 1986); L. Wohlpert, *The Unnatural Nature of Science* (Faber, 1992).
2. See, for example, B. Appleyard, *Understanding the Present* (Pan Books, 1992). I regret that something of this tone also creeps into the philosophical writings of M. Midgley, *Evolution as Religion* (Oxford University Press, 1988) and *Science as Salvation* (Routledge, 1992), justified though her criticisms of scientism undoubtedly are.
3. For a convenient critical summary of the thought of many leading twentieth century philosophers of science, see W. H. Newton-Smith, *The Rationality of Science* (Routledge and Kegan Paul, 1981).
4. B. Barnes, *Interests and Growth of Knowledge* (Routledge and Kegan Paul, 1977), p. 24.
5. For an account of this episode in particle physics, see J. C. Polkinghorne, *Rochester Roundabout* (Longman, 1989). The final chapter attempts a philosophical assessment.
6. T. Kuhn, *The Structure of Scientific Revolutions* (Chicago University Press, 1970).
7. See, for example, R. Rorty, *Philosophy and the Mirror of Nature* (Blackwell, 1980).
8. J. C. Polkinghorne, *One World* (SPCK, 1986), p. 20.
9. B. van Fraassen, *The Scientific Image* (Oxford University Press, 1980).
10. K. Popper, *Conjectures and Refutations* (Routledge and Kegan Paul, 1963); *The Logic of Scientific Discovery* (Hutchinson, revised edition, 1980).
11. I. Lakatos, *The Methodology of Research Programmes* (Cambridge University Press, 1978).
12. M. Polanyi, *Personal Knowledge* (Routledge and Kegan Paul, 1958).

13. Cf. J. Ziman, *Public Knowledge* (Cambridge University Press, 1968).
14. A counter-example might appear to be the conflicting interpretations of quantum theory (see, for example, J. C. Polkinghorne *The Quantum World* (Longman, 1984)). The issues here are partly physical (the nature of measurement) and partly metaphysical (the nature of causality).

3
Working together

Within a tradition

Although many scientists do not pay much attention to the history of their subject or display much concern with what sociologists might say about their activities, all scientists are inheritors of a tradition and members of a community. Beginning research students must make themselves masters of the literature that lies between the received understanding recorded in their undergraduate textbooks and the frontiers of knowledge of their chosen specialty. At the same time they begin to serve an apprenticeship to learning the way in which research is done.[1] This amounts to much more than acquiring necessary techniques, whether they be experimental or mathematical, for it involves the attitudes of commitment and relentless curiosity, fundamental to the quest for truth. A beginner faces the awkward question of how on Earth one can find out something new about the physical world, which has been subjected to generations of scientific interrogation? The answer is not just to be found internally, from whatever resources for speculation and ingenuity the student has been endowed with. It also comes from the external observation of how others, more experienced in the quest, pursue their investigations. One learns to do research by seeing others do it.

For many years, I was a senior member of a large research group of theoretical physicists in Cambridge. At the beginning of each academic year I used to give a talk, offering a little avuncular advice to those new research students who had just joined us. I used to say to them, with perfect honesty, that the most miserable year of my life had been my first year attempting to do research. I read many papers and mostly figured out what they were saying, but it seemed terribly difficult to find a new problem to try to tackle oneself. I had been a clever undergraduate, good at solving Tripos problems. Sometimes, of course, one had come across something rather difficult, and it might take a few days to sort it out. But one knew there was a solution to this undergraduate problem, which should be accessible, and in the end I usually found it. Now, as a research student, it was not only hard to think of new questions, but it was even harder to know if they were sensible questions, and hardest of all to guess whether their answers were in any way within the reach of persistent hard work. Months went by without any bright idea flashing into my mind. The adjustment to this slow rate of progress over a long timescale was very hard to make. Eventually one little niggling thought developed into something capable of yielding to a sustained attack and, in due course, elaborating this a little got me a Ph.D. and a Research Fellowship at Trinity College, Cambridge. At last I was under way and, though research continued to have its ups and downs, fallow periods of frustration alternating with more exciting periods of hot pursuit, I got used to the idea that one could, on average, sustain a degree of modest contribution to the subject.

My start in research had been made the more painful and difficult because of the circumstances obtaining at that time in Cambridge for postgraduate students in applied mathematics. There were then no departments in our faculty. We beginners sat alone in our college rooms, meeting only once a week for a seminar, held in a room in the Arts School, whose walls were garnished with a collection of nineteenth century plaster casts of mathematical models. Communal life was at the lowest possible ebb.

It was only when I went to the California Institute of Technology, to work as a Postdoctoral Fellow with Murray Gell-Mann, that I began to participate in the daily life of an active group of theoretical physicists, with much discussion and the opportunity to learn how the minds of experienced and successful persons approached the matter of doing research. It was a formative experience. A belated apprenticeship had begun, though as a smart young Fellow of Trinity I was, perhaps, only partially able to acknowledge this at the time. When I returned to Cambridge on the teaching staff, a few years later, the long-overdue departmentalization was beginning to happen. My colleagues and I fostered a group feeling and the generations of our research students enjoyed much discussion over coffee, in and out of our offices, and in fairly informal seminars. We all learnt from each other, for good apprentices soon become able to teach and stimulate their notional masters.

Handing on the torch

Helping research students get going requires delicate judgement. I usually tried to suggest a problem that had some interest and seemed do-able with the sort of techniques our group specialized in. This kind of apprentice piece could build up skill and confidence. Thereafter the best people (of whom we always had a significant number) tended to become self-propelled. Sometimes that would take them out of the group, to spend a year elsewhere where greater expertise relevant to their chosen interest was to be found. However, research is a chancy business. I remember talking to one student of ours, who has subsequently had a fine career, but who in the middle of his second year still had very little to show for his labours. He was understandably depressed, but fortunately agreed to stick it out a little longer. Within a few months the tide had turned for him. The most difficult students were those who were unwilling to fill the initial apprentice role and wanted to try something ambitious from the start. One or two of these were people who simply overestimated their innate ability but there were also one or two who were conscious of their considerable intellectual power and felt they should not condescend to initial engagement with the humdrum. In consequence, they achieved less than one would have hoped for.

In physics, theorists have the advantage over experimentalists in that their needs are simple: paper, a library, computing facilities, and a large wastepaper basket. In consequence, theoretical physicists enjoy a certain freedom to rove intellectually and to follow their fancy. Experimentalists, however, are tied to hardware, and apparatus can often prove intransigently difficult in its operation. In my old subject of elementary particle physics, the experiments are gigantic, lasting ten years from conception to analysis, spending millions of pounds in the process, employing literally hundreds of physicists. Here the novice will have to serve a more prosaic form of apprenticeship, assigned a specific task whose execution provides a small piece of the large empirical jigsaw being assembled. Nevertheless, the interest of the subject is such that bright young people are more than willing to serve in the experimental chain gangs.

Rivalry and collaboration

The execution of these experiments demands hours of time using very expensive accelerator facilities. There is a race of physics 'barons' who are the entrepreneurs of these great endeavours. Competition with rival groups for access to limited resources has ensured that most group leaders are larger-than-life figures, tough and uncompromising in the demands they make.[2] I remember a bright young postdoctoral researcher saying to me once, with some bitterness, that if you had a good experimental idea you had to persuade one of the barons to adopt it and, if he were an unusually nice fellow, he might even let you work on it. Theorists are luckier. They can go solo if they want to, though often good work comes out of small collaborations. I was fortunate in my career in having a very able younger colleague, Peter Landshoff, with whom I wrote many papers. We were sufficiently close to be able to work together, sufficiently different to complement each other. Together, I believe, we got more done than we could have managed separately.

The search for truth

The ambiguities of human nature and ambition are not absent from the community of scientists. Yet it is also a community of seekers after the truth. What motivates physicists to endure all the long hours of labour and weary frustration, inseparable from the pursuit of their discipline, is the desire to *understand* the physical world. That understanding is not gained just in some simple moment of individual illumination or through some inevitable, triumphal progress made by a happy few. It is a social endeavour, with all the false starts, blind alleys and illusory claims that characterize any form of human activity. My time in elementary particle physics coincided with the twenty-five years or so that it took us to discover a new level in the structure of matter.[3] When I was a research student in 1952, we believed that nuclear matter was composed of protons and neutrons. By the time I left the subject in 1979, we knew that protons and neutrons are themselves composites, made up of the celebrated quarks and gluons, and we had a theory, the so-called 'Standard Model', which seemed to describe pretty well what was going on. It would be easy, with hindsight, to give a 'Whig history' account of smooth progress (see p. 7), but in fact it was a long haul, with many dead ends, partial blindnesses, misleading 'good ideas', and even the occasional muddying of the waters by incorrect experimental results. I want to say two things about it all.

The first is that the story is one of two steps forward, one step back, of stuttering, lurching progression. The second is that nevertheless it is a story of progress – the dust did settle and a new and more profound understanding of the structure of the physical world resulted from it. Let me give an example to illustrate the point.

On and off in the 1960s, theorists toyed with the idea that two of the forces of nature, the weak nuclear force and the electromagnetic force, despite their striking differences, might nevertheless be manifestations of a single fundamental phenomenon. The hope was that they could be unified in a way similar to that by which the nineteenth century physicists had shown that the apparently different effects of electricity and magnetism could be interpreted by means of a single electromagnetic theory. There were a number of more or less obvious ways to try to bring off this twentieth century unification, but none of them could be made to work satisfactorily. Meaningless infinities kept on appearing in the calculations. Towards the end of the decade, however, Steven Weinberg and Abdus Salam independently hit on a subtle way in which the union could be achieved. The resulting 'electroweak theory' is one of the most important components of the Standard Model, for which its originators were rightly awarded shares in a Nobel Prize.

Yet the first announcement of their idea sank like a lead balloon. Weinberg's paper was *nowhere* referred to in the literature until three years after it appeared. Salam's was equally unnoticed, but he had published it in a rather obscure way, so that was less surprising.

Two factors contributed to this neglect of what was in fact a very important discovery. The theory required the existence of a phenomenon, called neutral currents, which the experimentalists thought they had shown did not occur. The investigation was tricky, for it depended on estimating the contamination ('background') due to other processes that could mimic the effect being looked for. In the 1960s the experimentalists believed that all the candidates they had discovered could be explained away as being spurious background effects. It is difficult not to believe that they reached this conclusion with some relief, because at that time the theorists were not 'in favour' of neutral currents anyway. The experimentalists seemed to be endorsing the prevalent orthodoxy. It is always hard for them to see what is not expected. The second difficulty was theoretical. Although Salam and Weinberg believed their theory was free from the problems that had plagued earlier attempts, they had not been able to prove that this was so. Maybe it was all another rather wild guess in an area where there had already been several wild and unsuccessful guesses. In any case, until the theory had been certified free from trouble, there were very few calculations that could be done with it with any confidence.

All this changed in the early 1970s. Better experiments, with better calculations of background effects to support them, showed that neutral currents were there after all, just in the way the electroweak theory predicted. At about the same time, a brilliant young Dutchman, Gerhard t'Hooft, showed that the Salam–Weinberg theory was mathematically well behaved. As a result of all this, there was an explosion of calculational activity, bringing with it empirically satisfactory results. Electroweak had arrived.

The tale is clearly not one of majestic inevitable intellectual advance. It is an untidy story of human endeavour, incorporating both insight and error. Yet it is a tale of real progress in understanding the physical world. The experiments of the 1970s were more accurate and better analysed than the experiments of the

1960s had been. There were objective grounds for preferring the former's conclusions. The new theory was elegant and well understood. The community of science had gained knowledge: there were neutral currents; electromagnetism and the weak nuclear force were aspects of a single underlying phenomenon. This was not just the result of a tacit agreement among the invisible college of high energy physicists to see things this way. The understanding gained was a discovery of what the physical world is actually like.[4]

Many philosophers of science are unwilling to accept that judgement. In Chapter 2, we saw that scientific theory is not logically entailed by experiment but rather it is read out of empirical knowledge by a personal act of creativity. That creative insight is sifted and endorsed within the truth-seeking community of scientists. Philosophers find it difficult to recognize how resistant nature is to yielding up its secrets, how very hard it is to discover a theory possessing economy, plausibility, and wide empirical adequacy. They suspect, somehow, that there are many such theories lying around, undisclosed because of the scientists' lazy acceptance of the first socially agreed version that comes their way. One of the reasons for rejecting that view lies in the activities of the scientific 'young Turks'.

Fame and fortune

In a field that is both fundamental and showing signs of opportunities for gaining further understanding (such as elementary particle physics was throughout my time in the subject), there is a considerable concentration of talent on a narrow front. Many very able young men and women are drawn in by the prospect of making their mark and gaining an international reputation. Far from being overawed by a socially controlled consensus, they are anxious to upset received wisdom and strike out in novel directions. If all this restless striving by people with great intellectual gifts does not turn up a plethora of new and plausible theories, it seems highly likely that this is because such new theories are just not there to be had. Theory-making is much more difficult than the philosophers are prepared to admit, and its successful conclusion carries with it a natural conviction of discovery.

While truthfulness in the understanding of reality is the aim of scientific endeavour, it would be impossible to deny that, for many scientists, fame is the spur. I have known some, including people of the greatest distinction, for whom matters of reputation are matters of indifference, but they are a comparatively rare breed. In many ways, the scientific community is somewhat isolated. Our culture makes very little attempt to take a serious interest in matters scientific, regarding them as arcane and inaccessible. The treatment of scientific discoveries in the press and other media is fitful, inadequate, often whimsical, seizing upon a doubtful or trivial incident and neglecting one of much greater significance. Books aiming to convey scientific ideas to a wider public are often given little or no attention in the review columns outside the scientific journals. The names of scientists of the greatest distinction are often completely unknown to the public. Paul Dirac was a theoretical physicist worthy to be mentioned in the same breath as Isaac Newton or James Clerk Maxwell, but even among educated people it is likely that his name would not be recognized by many, and very few would be able to say what were his outstanding discoveries.

It is to their peers, therefore, that scientists must look for the appreciation which it is a natural human desire to receive for one's achievements. When much talent is concentrated upon a limited range of problems, people are conscious that the winner will take all. Fame will fall on the first to publish. James Watson's frank and pushy account of how he and Francis Crick successfully beat the Californians to the great prize of the discovery of the structure of DNA[5] describes an atmosphere both of intellectual excitement and of competitive ambition. This description is widely recognizable by scientists in many other disciplines, who have made much less significant advances in their own work but who will have agonized equally about not being beaten at the post. Hence the great pressure to stake a claim to priority.

Different procedures have been set up in the different parts of the scientific community for the rapid establishment of the fact that X has made the breakthrough. Some disciplines make use of rapid-publication media, such as *Nature* or the various 'letters' journals. In my old subject of particle physics, the ink was scarcely dry on the manuscript before we despatched many duplicated 'preprint' copies to rival groups around the world. Today the electronic instancy of e-mail is devoted to similar rapid dissemination of results in the pursuit of priority. It all makes for a rather frantic life, both as despatcher and as recipient, with an inevitable degree of trigger-happiness resulting in occasional premature and unjustified claims. The loss of face and reputation resulting from such ill-judged efforts, however, serves to act as a powerful restraint on too much incautious behaviour.

These are the techniques for trying to get one's foot in the door of the Hall of Fame. Admittance to the building (or to its anterooms) comes with various institutionalized forms of recognition. Prime among these for scientists is, of course, the award of a Nobel Prize. Everyone recognizes the Laureates as men and women of high distinction, although there have been a few chosen whose claims have faded somewhat with time or even, in one or two rare cases, which have been shown to be ill founded.

The side-effects of a Nobel award are also significant, for prize-winners not only receive a substantial sum of money but they seem also to be granted by society a licence to receive a respectful hearing on topics lying far outside their acknowledged expertise. A few have chosen to take extensive advantage of this opportunity.

For a long time there was a rather sparing policy in Stockholm as far as rewarding outstanding individuals was concerned. Rutherford was awarded a Chemistry Prize for his work on radioactivity, but all his later great discoveries (such as the existence of the atomic nucleus) were unrewarded. He never received the Prize for physics. Einstein received his Prize for his work on the photo-electric effect; the discoveries of special and general relativity were not the subject of further awards. However, in 1972 John Bardeen, who had shared the Physics Prize in 1956 for the development of the transistor, was awarded a second share of the Prize for joint work elucidating the origin of superconductivity. What had previously been supposed to be a convention (no second participation in the same subject Prize[6]) had been breached. I had the clear impression that this event produced a distinct uneasiness in the highest echelons of the scientific community. To emerge from the throng as absolutely pre-eminent, one Prize was no longer quite enough! Yet so far only Fred Sanger (two Chemistry Prizes) has joined Bardeen in the 'double first' super-elite.

Most scientists are not going to spend much time worrying in late October/early November whether that magic telegram from Stockholm is going to arrive. It is pretty easy in science to assign people, including oneself, to broadly correct bands of achievement. One knows roughly speaking where one is and what recognition one might hope for. Real generators of widespread anxiety, therefore, are going to be those awards that are at a high enough level to signify a satisfying achievement, but not so high that they are realistically beyond one's grasp. An example in the Commonwealth community would be election as a Fellow of the Royal Society. You certainly do not have to be a genius to get in, but selection does signify an honourable degree of professional attainment. Because the pyramid of talent widens as it goes down, many more scientists will worry about getting into 'the Royal' (to use the somewhat cosy way of referring to it, popular among its Fellows) than about getting a Nobel Prize, and many more will be in that unsettling borderland where one might have got in but one was not lucky enough to do so. I have to admit that the ambition to be an FRS was a potent and disturbing element in my scientific life for a good number of years. If you had put to me some curious scheme by which my election would have been assisted by the murder of my grandmother, I would certainly have declined, but there would have been a perceptible pause for mental struggle before I did so.

One of the British diseases is our love of classifying people, our donnish delight in discriminating the α^- from the β^+. This makes the possession of the initials FRS after one's name particularly desirable and particularly productive of other agreeable benefits. I very much doubt whether I would have been elected President of Queens' College in 1989 if I had not already been elected FRS in 1974. There are a good many positions in the British academic world for which FRS (or its arts equivalent, FBA) is effectively a *sine qua non*. Yet I must record one sociological curiosity. British society is much given to proclaiming status (school or university, elite club, Army regiment) by the wearing of an exclusive tie. It is odd that the Royal has never produced a unique tie for its Fellows to wear.

The conference circuit

One setting in which claims for priority (and hence for recognition) could be made is represented by international conferences at which results are reported and discussed. My own former subject of particle physics has had a series of this kind, which was started in 1950 and which still continues. They are called Rochester Conferences because they were started by Robert Marshak at Rochester University, where the first seven gatherings were held before the Conference became a world traveller. The changing character of the Conferences over the years has corresponded to the sociological development of the high energy physics community.[7] The original meetings were small, informal, run on a shoestring as far as finance and administration were concerned. The first one I attended was in 1957. We were still a small enough gathering for people to be able to speak about their own work, one by one. The timetable for the speakers in the theoretical sessions was written on a blackboard and amended by the chairman from time to time as self-important speakers failed to observe their allotted ration of time. It was friendly and good fun, a relaxed 'town meeting' of the international intellectual village populated by the high energy physicists. This neighbourly informality could not survive the rapid growth of the number of scientists working in the subject.

Far too many papers were eventually submitted to the Conference for each author to be able to have his or her own slot. A 'rapporteur' system grew up, by which senior physicists were allotted topics. They dealt with all the papers in their assigned area and then produced sifted and digested assessments of what progress had been made since the last Conference. The approbation (or neglect) of the relevant rapporteur became an important element in the generation of your reputation. On the whole, those chosen for the task tried to do a careful and responsible job. Some, of course, were so confident of their idiosyncratic judgement that they felt no need to give attention to those whose work lay outside their self-selected orthodoxy. Some (the worst of all) declined to exercise judgement at all and just attempted a high speed rattling-off of all that had been submitted to them.

Such a system could not leave everyone happy with its results. The discussion period at the end of the report, meant for the further elucidation of the issues presented, tended to become hijacked by those who wanted to insert their own idea which the rapporteur had strangely omitted to mention. The cant phrase 'For the sake of completeness . . .' became the oft-repeated introductory remark in such attempts at self-advertisement. I doubt whether this technique was really of much help to those who felt they had suffered unjust neglect. Its self-promotional aspect was too transparent.

As the high energy physics community continued to expand, these large conferences, attempting a synoptic view of the whole subject began to loose their charm and some of their utility. A more favoured enterprise became the elitist kind of gathering represented by a 'workshop', a meeting drawing together a small number of leading experts in a well-defined field and permitting more relaxed and effective interchange of ideas. Certainly, high energy physics continues to be a personal activity, drawing vitality from the face-to-face encounters of those engaged in it. E-mail may supplement conversation but it will never replace it.

The ordinary and the extraordinary

A Whig history of science would take account of only the great men and women, and even for them, would notice only their great discoveries. That would be a distortion of the truth for two reasons.

The first point is that we members of the army of honest toilers also have our role to play, as exploiters and developers of great ideas and as contributors to the atmosphere of inquiry in which the great ideas can first emerge. Of course we shall disappear from collective memory – for unless you are a Newton or a Maxwell, science is ultimately an anonymous activity – but we shall leave behind our modest deposit of technique and understanding. Our memorial, for a while at least, will be our students, for it is from the average professor of physics that most of the next generation, including its geniuses, will get their initiation into research and, one hopes, the kindling of their love for the subject.

The second distortion present in Whig history relates to the great scientists themselves. We have to be aware of their limitations as well as their genius. It often seems that their greatness lies in seeing with signal clarity the one thing needful for the contemporary advance of science, and their pursuing that point with single-minded persistence. They ask the right question at the right time. Thus Einstein had the insight to question the classical concept of simultaneity and this led him to the theory of special relativity, to which Poincaré and Lorentz had come so close in terms of formalism but yet were so distant in terms of concept. They had the right equations but they had not found the right meaning. A few years later Einstein saw that the equivalence principle held out the prospect of a new approach to gravity. This principle asserted the universal equivalence of gravitational mass and inertial mass, that is to say that exactly the same quantity determined both how a body was influenced by gravity and also how it resisted having its state of motion changed by gravity. Double one, and you doubled the other. The resulting effect remained the same. This implied that bodies would move along paths in a gravitational field that were independent of the amount of matter they contained, which in turn implied that their motion could be represented geometrically, as an effect of the curvature of space itself. From this insight sprang the modern theory of gravity, successor after more than 200 years of splendid physics to the theory of Sir Isaac Newton himself. It was a wonderful achievement, effectively accomplished, as far as Einstein was concerned, by 1920.

Einstein was then 41. He spent the remaining 35 years of his life principally in endeavours that were essentially fruitless. He felt that the next step was to unify gravity and electromagnetism (the only two basic forces of nature then clearly recognized) and his intuition was that geometry should once again provide the key. Obviously, this would be more difficult, since charged particles behave quite differently from uncharged particles in the presence of electromagnetic fields. There is no longer a universal form of behaviour. Brilliant though he was, Einstein was in the position of a general seeking to win new battles by following the tactics of the last war. That particular kind of geometrical insight was no longer the one thing needful, and someone has said that Einstein would have been better employed in the last quarter century of his life in going fishing. Yet in fairness, we must remember that this period included his fundamental work with Podolsky and Rosen, which lead eventually to the recognition of non-local effects in quantum theory[8] (but not as Einstein had hoped, the discrediting of quantum theory).

This story is by no means an isolated example. In fact, it recounts a common fate of the very distinguished. Werner Heisenberg was one of the greatest physicists of the twentieth century. His contributions ranged from his fundamental role in proposing modern quantum theory and eliciting its properties, to his important contributions to theories of the magnetic properties of solids and the turbulent motion of fluids. Everything he turned to he seemed to illuminate, until the closing years of his life. He then succumbed to the hubris of supposing that he could, out of his head, write down a fundamental non-linear spinor equation from which one could read out all the basic properties of matter. It is even said that Heisenberg advised the German government funding agencies that it would be better to spend money on solving his equation than on building accelerators to investigate the actual properties of matter! It was a final bid for glory, but it proved to be ill judged. In the attempt to exploit his idea, Heisenberg decided that one should make use of a dubious technical device, which is called an indefinite metric. This is a dangerous notion, because it introduces into the formalism the hazard of calculating negative probabilities, which are, of course, nonsensical. Heisenberg had attempted to recruit the assistance of his old friend, Wolfgang Pauli, well known as an acerbic critic of doubtful ideas. Pauli resisted at first, but apparently Heisenberg's persistence wore him down, for eventually he joined the project. His adherence did not last long, and his subsequent denunciations were correspondingly the more vehement because of his temporary loss of judgement. It all culminated in a highly public repudiation of Heisenberg by Pauli at a session of the 1958 'Rochester' Conference, held that year in Geneva. Heisenberg was the speaker, Pauli, the chairman. Describing the scene, I wrote

Heisenberg was continually interrupted by Pauli, who wagged a reproving finger whilst uttering such remarks as 'mathematically objectionable', 'This I discussed [and rejected] in April and I wonder that you again repeat it all' and, most frequently, 'no credits for the future'. It was a scene at once farcical and sad. Justification lay with the sceptical Pauli but Heisenberg was one of the greatest physicists of the twentieth century who should have been able to enjoy a more dignified close to his career.[9]

It is hard for the great to grow old gracefully in physics. I do not want to pretend, however, that it cannot be done. Sometimes it happens without being recognized at the time. When I was a young physicist, strong interactions, involving pions and nucleons, seemed to be where the action was. Our venerable and venerated Cambridge colleague, Paul Dirac, seemed oblivious of all this contemporary excitement. He continued to think and write about fundamental properties of relativistic quantum mechanics. To us would-be young Turks, it all seemed a little quaint and passé. Only much later would we come to recognize the importance of Dirac's work on such topics as monopoles, a pioneering encounter with those topological aspects of quantum field theory that have assumed great significance in recent thinking.

It is perhaps a little easier for the honest toilers to grow old. They tend to be more flexible than their distinguished contemporaries, broader because they are shallower in their insight. There is an opportunistic element among them; they do what they can. Physics is for them the art of the possible. They do not suppose that they have seen the one thing needful, but they are prepared to try their hand at whatever looks promising. Bumming around in the world of physics, as one might say, enables them to keep on the road a little longer.

Changing direction

But it cannot last for ever. This is particularly so for theoretical physicists. All mathematically based disciplines call for a certain flexibility of mind, which most of us lose as we get into middle age. I had long thought for myself that I would not stay in theoretical physics all my working life. I had seen too many of my seniors get somewhat miserable as the subject moved away from them. As my fiftieth birthday approached, I was conscious of becoming more and more breathless as I ran to stand still in the ever-changing world of high energy physics. Such rapid advances were stimulating when one was young. They became exhausting as one grew older. I felt that I had done my little bit for physics and that the time had come to do something else. What that something else was is another story.

I always want to emphasize that I did not leave physics because I was in any way disillusioned or discontent with it. I had greatly enjoyed my time in the subject and I retain an affection for it and an interest in its advance, even if I now can learn about the latter only in a fairly broad brush way. (Technical mastery can be achieved only by full-time commitment.) So my memories are grateful. That gratitude refers not only to the intellectual pleasures that living through the quarks-and-gluons revolution afforded me but also to the many friends I made throughout the world-wide intellectual village of high energy physics. My strongest nostalgic twinges for the old times relate to the companionship of my colleagues, which I enjoyed over the period of more than twenty-five years that I spent in the subject. It was a privilege and a pleasure to have been able to spend that time in the community of scientists.

NOTES

1. M. Polanyi, *Personal Knowledge* (Routledge and Kegan Paul, 1958), especially ch. 7.
2. Cf. P. Galison, *How Experiments End* (University of Chicago Press, 1987).
3. See J. C. Polkinghorne, *Rochester Roundabout* (Longman, 1989).
4. For a detailed defence of this assessment see *ibid.*, pp. 166–8.
5. J. D. Watson, *The Double Helix* (Weidenfeld and Nicholson, 1968).
6. There have been a small number of recipients of two different Prizes.
7. For one man's account see Polkinghorne, *Rochester*.
8. See, for example, J. C. Polkinghorne, *The Quantum World* (Longman, 1984), ch. 7.
9. Polkinghorne, *Rochester*, p. 77.

4

Memoirs of the great

I have already declined to take a Whig view of the history of scientific progress and I have spoken on behalf of the army of honest toilers who make their modest contributions to the steady, if somewhat erratic, advance of knowledge. Science is an activity pursued within a community. Yet, having said all that, one must admit the vital role of the truly great, those who by their seminal discoveries open up new domains of investigation. Beneath the superficially egalitarian style of much of scientific society, the elite are accorded an intrinsic respect. As persons they are the focus of interest, and stories circulate about them. These tales are neither awe struck hagiography nor guttersnipe gossip. Often they focus on the idiosyncrasies of the person concerned, but in a way that is affectionate rather than demeaning. We are proud of our great men and women. One of the pleasures of long service in a particular branch of science is that one becomes acquainted with the leading figures.

Paul Dirac

Perhaps the greatest theoretical physicist I have know was Paul Dirac, one of the founding fathers of quantum theory and the greatest British theorist of this century. I had just about heard of him before I came up to Cambridge in 1949 to read for the Mathematical Tripos. Dirac did not lecture to junior undergraduates but one day in the foyer of the Arts School (where all the maths lectures were given) I saw a tall figure whose hair curled slightly at the edges and who was clearly someone of distinction. For some reason, the image of a French poet came into my mind, but I dismissed it and felt sure this person must be Paul Dirac. I was

right. In my final undergraduate year I went to his celebrated
course on the principles of quantum mechanics – learning the subject from the horse's mouth, so to speak. Dirac was an exceptionally
clear lecturer and one was caught up into an ecstatic realization of
the beauty of fundamental physics. No rhetorical tricks were
employed by the lecturer to produce this effect. It resulted from
the quality of the thought and the consequent unfolding of an intellectual theme as apparently inevitable and majestic in its elaboration
as the development of a Bach fugue. As well as undergraduates,
the audience always contained some senior visitors to Cambridge,
for even those who had spent many years working in quantum
physics rightly wished to hear it expounded by one of its masters.
One of the most impressive features of the lectures was the modesty
and self-effacement of the lecturer. Dirac had played a leading role
in the elucidation of non-relativistic quantum theory and he was
the founder of relativistic quantum mechanics. Yet there was never
any reference to, let alone emphasis on, his own achievements.

I always thought of Paul Dirac as being a kind of scientific saint.
There was an austerity and singleness of mind about him, a purity
of vision and humbleness of heart, which made one think naturally
in those terms. Many stories clustered around him and they were
regularly retailed in the physics community. They all centred on
a certain logical directness and simplicity that characterized, not
only his scientific work, but also his comparatively brief indul-
gences in everyday conversation. The quintessential Dirac story
concerns the aftermath of a lecture in which he had been
expounding his latest ideas. It had made demands upon his audience and at the end someone got up and said he had not quite
followed how Professor Dirac had derived such and such an equation. A long silence followed. Eventually the chairman said
'Professor Dirac, are you not going to answer Dr X's question?'.
There came the crisp and logical response, 'That was a statement,
not a question'. On almost anyone else's lips, that would have been
a putdown. On Dirac's lips, it was just a straightforward remark.
The same was true of a comment I once heard him make in the
Cavendish Laboratory tea room. Conversation had got round to
the physics of the 1930s, when many interesting developments took
place, exploiting the possibilities made available by the discovery

of quantum theory in the mid 1920s. Dirac simply said about that later fruitful era that, 'It was a time when second-rate men did first-rate work.' It was a matter-of-fact comment by an unquestionably first-rate man.

Abdus Salam

When I started research, my first supervisor was Nick Kemmer, a man of great charm who had made extremely important discoveries just before the Second World War. However, after a year, he left Cambridge to take up a professorship at Edinburgh and I was them taken on by his successor, the Pakistani theoretical physicist, Abdus Salam. He was a prolific generator of ideas. Salam has about him an air of uncontrolled intellectual fertility. Some of his ideas have been very good indeed – he is a Nobel laureate – but some of them have been, shall we say, less inspired. People with this kind of gift and temperament – I think Fred Hoyle is another – function best when they have in their neighbourhood a strong and more cautious personality, able to say 'Wait a minute' and to act as an intellectual filter and scientific conscience. Paul Matthews was for many years a partner of this kind in relation to Salam. Oddly enough, this fecundity did not impinge on me very much when I was a research student. Abdus mostly left me to pursue what interested me. Later in life though, he was to have a group of devoted postdoctorals (mostly located in his International Institute at Trieste) who busied themselves with the latest Salam research programme. We used to call them, in a friendly way, 'the chain gang'.

Salam's exuberance of intellect extended to his lecturing style. At an international conference, people would always be anxious to learn what he was thinking about. His innovative mind would be in gear until the last moment and he would turn up with a disreputable collection of casually scribbled transparencies for projection overhead in the course of his talk. One by one they would have been baffling enough, but it was Abdus's habit to put the next one on before he had taken the last one off. One got an impression of intellectual excitement, if not always a clear notion of exactly what the excitement was about.

A very attractive feature of Salam's career has been his desire to assist bright young physicists from the developing world. He had shown his own intellectual power quite precociously in his native Pakistan, then came to Cambridge to do the Tripos and a Ph.D. Thereafter he was a prominent member of the international world of theoretical physics. Salam was conscious of two things. One was the need for people to be in constant touch with new developments in fields that were highly competitive and rapidly changing. The other was the need for developing countries not to lose their brightest talent to permanent positions in the developed world. To solve both problems at one blow, Abdus Salam worked to found and fund the International Centre for Theoretical Physics at Trieste. Bright young people could be given Fellowships stretching over a period of years, which enabled them to spend three months a year at Trieste (keeping in touch with international research) and nine months at home (contributing to their country's university system). It is an imaginative and valuable scheme.

Murray Gell-Mann

I spent my own first postdoctoral year working at the California Institute of Technology (Caltech) with Murray Gell-Mann. I have already spoken of what a formative experience that was for me in my physics career. Gell-Mann was only a few years older than I was, but he already had an established reputation as a great theoretical physicist. For twenty years he was to be the leading figure in the competitive and talented world of elementary particle theory.

Murray is not only a great scientist. He is also a powerful personality. I learnt early on how one can survive and profit from being in orbit around so brilliant a star. One must not come too close. It is one thing to bathe in the light and warmth emitted; it is another, to be burnt to a frazzle. In his immediate presence, Gell-Mann's strength of character and his quickness of intellect meant that his thought patterns were imposed upon you and, of course, you could never out-think him. The best one could manage was to limp along breathless in his wake. (Murray once told me that he had only had one junior colleague who he thought approached him in intellectual power. Interestingly, that person has not, by the very highest standards, had an outstanding career.) If one were to have a life of one's own, one had to catch sight of where the action was, glimpse some modest aspect that might be amenable to one's own limited skills and ability, and then retreat to carry on with it by oneself.

One of Gell-Mann's outstanding talents is to have a nose for areas of work that are both significant and open to exploitation. During his heyday, the repeated question in the theoretical community was 'What is Murray up to now?'. His greatest achievement has been his eminent role in establishing the quark structure of matter. It all started with the discovery of a new quantum number; that is to say a quantity which controls what goes on in a domain of physics and corresponds, therefore, to an intrinsic property of the entities involved. This particular discovery explained some peculiar properties of the reactions involved, which differed markedly from natural prior expectation. Consequently Gell-Mann coined the term 'strangeness' for this new quantum number. Not only has this joke remained fossilized in the terminology of elementary particle physics, but it has also encouraged a rash of similar facetious namings, such as 'charm', 'bottom' and 'top'. It is an aspect of Murray's influence on my old subject for which I feel distinctly less than total admiration.

The next Gell-Mann physics joke was more sophisticated. He had discerned a significant pattern in the structure of elementary particles (SU3, for the initiated) among its consequences being the existence of octets of particles with associated properties. The paper enunciating this made punning reference to Buddhism in calling this insight 'the Eightfold Way'. A curiosity of this particular incident is that the paper reporting these results never appeared in print in an archive journal such as *Physical Review*. It had circulated so widely as a preprint – and Murray had received such extensive recognition for it – that he never bothered with conventional publication.

Most learned of all the Gell-Mann jokes was the naming of the constituents that might lie behind the discerned patterns of the Eightfold Way. 'Quarks' was taken from James Joyce's *Finnegan's Wake* ('Three quarks for Muster Mark') and it reflected

Gell-Mann's considerable interest in matters linguistic. He has always been free in making available in a didactic fashion his extensive knowledge of languages.

Quarks explained the observed patterns of particles in a formal, mathematical way, but it did not necessarily follow that they actually existed as physical constituents. Current belief in quark reality stems from certain experiments in which projectiles appear to 'bounce back' after collision with point-like quarks inside the target particles (see pp. 9–10), but it is still the case that no one has seen an indubitable instance of an isolated quark. Contemporary orthodoxy explains this by supposing the quarks to be 'confined'; that is to say so tightly bound within particles that no impact is strong enough to expel one of them. For a long time, the matter of the reality of quarks seemed unclear. Throughout this period, Murray Gell-Mann was soberly cautious. He often spoke about quarks, but he was careful to qualify this talk by referring to them as 'presumably mathematical'. Reflecting on this usage, I once wrote:

> I always considered this to be a coded message. It seemed to say, 'If quarks are not found, remember I never said they would be; if they are found, remember I thought of them first.'[1]

I was perturbed recently to read my words quoted by James Gleick, with the comment that they represented a common view in the physics community but one which 'For Gell-Mann . . . became a permanent source of bitterness'.[2] I would be sorry indeed if that were so. It would arise from a failure to recognize that there is an affectionate element in the way physicists note the foibles of those whom they admire. After all, the photograph of Murray that appears in my book from which that quotation is drawn is subtitled 'a hero of our tale'.

Gell-Mann's strong personality reflected itself in a robust way with questioners, particularly those who had rushed in where angels fear to tread. A favourite technique with an ill-judged interlocutor was to pause, as if searching his mind to find what on Earth it might be that this rash person could have been thinking about, and then to say 'Oh – you mean . . .', followed by some extremely dismissive formulation of the foolish position taken up by the questioner. Gell-Mann can be devastatingly formidable. It is not surprising that when he encountered a jaguarundi in the forests of Central America, it was the jungle cat that turned aside.[3]

Salam was an outpourer of ideas and he did not really mind whether you lost some, provided you also occasionally won some. Gell-Mann, however, was very careful not to make public mistakes. His theories have stood the test of time, though in one or two cases there have been modest revaluations of the significance attached to them. If you had asked me in 1970 what was Gell-Mann's greatest contribution to physics I would have been tempted to reply 'current algebra'. This was a very clever idea that abstracted certain algebraic properties from simple quark models and suggested that they might also be characteristic of the (then unknown) full theory of quarks. These relationships could be manipulated to give testable results for experimentally measurable quantities. It was an ingenious and imaginative idea, and it led to some very interesting physics. However, with hindsight it now looks less significant than the original quark ideas from which it sprang. That is because unexpected advances in our understanding of basic quantum field theory have led to the possibility of constructing a plausible, and at least partly usable, theory of quarks (quantum chromodynamics), an achievement that seemed infinitely remote in the 1960s, a time of maximal disillusionment with field theoretic ideas.

Richard Feynman

When I was at Caltech I got to know another great theoretical physicist, Richard Feynman. He and Gell-Mann were about as different from each other as they could be. Murray was the intellectual polymath; Dick cultivated the image of a fun-loving kid from Queens, New York, who just happened to be an ace physicist. It was funny to listen to them talking. Gell-Mann always pronounced foreign words in the correct native fashion. Feynman would affect not to understand what he was referring to. Even so simple a word as 'Moscow' could be the source of sparring and simulated

misunderstanding. In one of his autobiographical writings, Feynman quotes some remarks of his father about knowing a bird's
name:

> You can know the name of that bird in all the languages of the world,
> but when you've finished, you'll know absolutely nothing whatever
> about the bird. You'll only know about humans in different places
> and what they call the bird. So let's look at the bird and see what
> it's *doing* – that's what counts.[4]

Physicists reading this anecdote cannot help thinking of their
linguistically talented, ornithologically expert, colleague, Murray
Gell-Mann.

Many people seem to have found Feynman's first memoir about
himself, *Surely You Must Be Joking, Mr Feynman?*[5], to be an
entrancing read. I must confess that I was chilled by it. Its surface
message is the projection of the image of the unconventional, un-
inhibited, scourge of the bourgeois intellectuals, who also just happened to have a Nobel Prize for physics; its clear subtext is, 'I'm
cleverer than anyone else and here are a hundred anecdotes to
prove it.' In print, as in life, Feynman's restless jokiness soon
becomes wearisome. The persona takes over the man.[6] He tells us
that he did not really want to accept the Nobel Prize.[7] I do not
for one moment believe that to be true. Friends who were at Caltech when he got the award have told me what a pleasure and
satisfaction it was to him (and why not?). Certainly this seemed
reflected in a new burst of physics activity that followed the award,
succeeding a somewhat doldrum period that had preceded it.

Feynman and Gell-Mann did not only differ in personality.
They differed also in the kind of physics they pioneered. Dick
Feynman's greatest discovery was a wonderfully insightful and
accessible way of doing calculations, universally known as the
Feynman diagram technique. It has proved astonishingly fruitful,
not only in the master's own hands but also as a weapon in the
armoury of the army of honest toilers. I write with some feeling
for I (like many others) spent the greater part of my physics career
working with Fenyman diagrams and eliciting some of their rich
and instructive properties. Their invention was surely the most
fertile technical gift to theoretical physics in the last fifty years.

Yet even the great, it seems, are not satisfied with their attainments. Feynman longed to discover a law of nature (as Gell-Mann had done with his strangeness ideas and the theory of quarks). In 1957 he thought he had done so, by producing a new theory of weak interactions (technically, V–A theory). It was certainly an important idea, which almost anyone would have been proud to have had. Yet it was also an idea that was very much in the air at the time, and in fact it was simultaneously thought of by several other people (including Gell-Mann – he and Feynman eventually wrote a joint paper). I would not place it towards the top of Feynman's achievements but it is rather touching to read his own response to this piece of work. He said 'There was a moment when I knew how nature worked. It had elegance and beauty. The goddam thing was gleaming'.[8]

Feynman had a deep intuitive understanding of physical phenomena of all kinds. This manifested itself in many ways, ranging from the celebrated *Feynman Lectures in Physics*[9] (characteristically written up by someone else from Feynman's performances – he had a great sense of the theatrical, but he was an actor rather than a playwright) to a lovely series of short programmes recorded for BBC television in which he took everyday phenomena and unpacked the interesting physics that lay behind them.

Dick Feynman was a restless genius, caught up in the toils of an idiosyncratic image of his own making. It is interesting that the self-effacing Paul Dirac was his great scientific hero. Feynman once found himself sitting next to Dirac on the High Table at St John's College in Cambridge. For once he was lost for words. After a long silence, Feynman blurted out 'It must have been wonderful to discover that equation [of the electron]'; Dirac said 'Yes. One must look for beautiful equations'. Another long silence followed. Then Dirac said 'Are you looking for an equation?'. Feynman spent his life looking for an equation. Though he discovered many beautiful solutions, that particular Holy Grail of physics eluded him.

Stephen Hawking

I spent most of my professional life as a theoretical physicist working in the Department of Applied Mathematics and Theoretical Physics at Cambridge (DAMTP for short and pronounced acronymically with the p and t interchanged). A very remarkable colleague was Stephen Hawking. I can recall him coming to us as a graduate student to work with Fred Hoyle. Hawking was then physically somewhat maladroit. Most of us did not know that this was the onset of motor neurone disease, against which Stephen has fought so doggedly and with such amazing fortitude and success. He is a person of iron will and determination and this strength of character has fortunately kept him alive long beyond the initial expectations of the physicians. If that were all to his story, it would by itself make him a remarkable man but, of course, he is also a very distinguished theoretical physicist. The idea that quantum theory makes black holes radiate thermodynamically, with a temperature decreasing with the mass of the black hole and an entropy proportional to the area of its event horizon, is a deep and wonderful concept, uniting quantum mechanics, relativity and thermodynamics in a most profound and satisfying synthesis. It is an idea that is clearly of Nobel calibre, but whether Hawking will ever receive the Prize is more uncertain. The Swedish Academy is always cautious and it demands experimental endorsement of theoretical ideas. In the case of black holes (even if glowing a bit through Hawking radiation) such confirmation is not easy to obtain. But I do not think anyone doubts that Hawking is right in his proposal.

Who would have guessed ten years ago that Stephen Hawking would add to his list of achievements that of being the author of a book that has been the most remarkable phenomenon of twentieth century scientific publishing? The millions of copies sold of *A Brief History of Time*[10] make it the scientific best-seller of the century. In fact, it is all a bit of a puzzle. Why did so many people buy it? What have they managed to get out of it? Among the chattering classes it has become fashionably acceptable to say that one could not get 'beyond page x' (where x is a small number). Yet one has to be able to say that one had at least tried it out.

It seems to me that among the non-scientific there is a good deal of suppressed uneasiness in people's attitudes to *Brief History*. This arises partly from a feeling that here is the answer to the riddle of the universe – even if one cannot quite follow what it is – and so one ought in one's ignorance to pay some sort of respect to that, without quite knowing how to do so. But, of course, *Brief History* is not the answer to the Great Question of existence. We do not even have a proper theory of quantum gravity, since the two great discoveries of twentieth century physics, quantum theory and general relativity, still remain imperfectly reconciled with each other. Thus, the speculations about quantum cosmology that occupy the last quarter of the book, and which are Hawking's personal, and by no means undisputed, approach to the question of the nature of the very early universe, though scientifically interesting and stimulating, are in no sense definitive. Even if they were, they would be of only marginal relevance to the great questions of metaphysics.[11] When Hawking chooses to talk of philosophical or theological issues, it is very easy to convict him of *naïveté*. An entirely justified admiration for the author should not lead us to canonize his writings.

Among the scientific readership, different forms of uneasiness lie beneath the surface. It is impossible to read (or review) the book without remembering the immense cost of authorship involved. What we can scribble off without effort, must be laboriously assembled by Stephen Hawking using his electronic devices. A paragraph of *Brief History* must have involved as much effort as a chapter from an unhandicapped writer. That realization has made it very difficult to say what I believe is in fact that case, that though *Brief History* is clearly set out with some nice humorous touches, it is not in itself at all a remarkable book. Written by Joe Bloggs, it would doubtless have sold a few thousand copies and that would have been that.

Such a thought induces an even less acceptable thought in the minds of those of us who have tried our hands at writing about science for the general educated public. It is, simply, envy. We have sold in our thousands, or at best tens of thousands, while Hawking has sold in his millions. Yet we have the sneaking feeling that if the public had tried us, they might have got beyond page *x* with some profit.

Finally, the most difficult thing of all to say is this: Stephen Hawking is probably the best-known living scientist. He is certainly a great physicist, in the Nobel class, and we pygmies salute his giant greatness. Yet, I do not find it easy to reach for phrases, so popular in the public mind, like 'the second Einstein'. One is already talking in superlatives when one refers to Nobel calibre, but comparisons with Einstein are something yet again. Here one is talking of a very small set of people (Newton, Maxwell, Einstein . . .) who have transformed our understanding of the physical world. Nobel Prizes are awarded every year; people of this stratospheric kind appear only once or twice a century. One of the really disturbing, and saddening, aspects of *Brief History* was the curious, inadequate, pairs of pages about Galileo, Newton and Einstein, inserted at the end. If they were there as an invitation to nominate the next member of the apostolic succession in physics, great care would be needed in suggesting the rightful name.

This rather miscellaneous collection of anecdotal reminiscences and personal judgements has had a serious purpose. It is to convey something of the feel of science as an activity of persons, conducted in a community where the great leaders of thought are, perhaps, more readily recognized and acknowledged than in any other community, and where they are valued, not only for their discoveries but also for their characters, varied, talented and limited, as all human personalities must be.

NOTES

1. J. C. Polkinghorne, *Rochester Roundabout* (Longman, 1989), p. 110.
2. J. Gleick, *Genius* (Pantheon, 1992), p. 391. M. Gell-Mann (*The Quark and the Jaguar*, Little, Brown and Company, 1994, p. 182) has recently explained that by 'mathematical' he meant 'confined', and by 'real', unconfined. Such idiosyncratic linguistic usage was bound to lead to confusion.
3. Gell-Mann, *Quark and Jaguar*, pp. 6–7.
4. R. P. Feynman, *What Do You Care What Other People Think?* (Norton, 1988), pp. 13–14.
5. R. P. Feynman, *Surely You're Joking, Mr Feynman!* (Norton, 1985).
6. One of the great merits of Gleick's biography is that it penetrates beneath the persona.
7. Feynman, *Joking*, pp. 305–6.
8. Quoted in Gleick, *Genius*, pp. 338–9.
9. R. P. Feynman, R. F. Leighton, and M. Sands, *The Feynman Lectures in Physics* (Addison-Wesley, 1963), three volumes.
10. S. W. Hawking, *A Brief History of Time* (Bantam, 1988).

11. See J. C. Polkinghorne, *Science and Christian Belief* (SPCK, 1994; also published as *The Faith of a Physicist*, Princeton University Press, 1994), p. 73.

5

What happened to the human mind?

The cumulative discoveries of science concerning the structure and history of the physical world comprise one of the great intellectual triumphs of humankind. It is a remarkable fact that our minds have proved capable not just of coping with everyday experience but also of penetrating the secrets of the subatomic world and of exploring the nature of the vast space of our universe, with its trillions of stars. Quantum theory and cosmology are among the highest achievements of twentieth century thought. Yet where in that world described by science can we locate the mind itself? The lifeless landscape that is the scientific domain, that location of transactions of energy moving between constituent bits and pieces, seems to afford no obvious home for the mental. Paradoxically, scientific thinking has allowed mind to slip through the wide meshes of its rational net. All that we can learn directly from science relates to the (doubtless very important) discoveries of neuroscience about the structure and activity of the brain. Yet there is an ugly big ditch yawning between scientific accounts of the firings of neural networks, however sophisticated such talk may be, and the simplest mental experience of perceiving a patch of pink.

Bridging the gap

At work in that ditch are the psychologists, whose investigations of such topics as the functioning of human memory provide a source of understanding intermediate between neurophysiology and the phenomenology of mental experience. I would also wish to consider seriously the insights offered by the depth psychologists. Freud and Jung and their successors are scarcely unanimous in the maps that they offer of the unconscious psyche, but that there is a dimension to our minds more profound than that of the conscious rational ego seems clear to me. It is a common scientific experience that a period of intense but unsuccessful engagement with a problem, when followed by a fallow period in which unconscious processes are presumably at work, can yield the sudden appearance into consciousness of the solution, fully formed, as it were, by subterranean process. The story of Henri Poincaré grasping the complete answer to a deep mathematical problem with which he had fruitlessly struggled for many months, at the very moment of stepping onto a bus to go away on an expedition, is a striking example of a phenomenon that many of us have experienced in lesser ways.

The ultimate bridging of the gap between mind and brain is the task of metaphysics. Its conceptual constructions will rest upon the empirical foundations provided by scientific insight, but they will no more be determined by science alone than the layout of the foundations of a house will fix the shape of the edifice built upon them. There must be consonance – they must fit together – but there is no entailment. If we are to make progress we must address ourselves to the philosophy of mind. I do so conscious of my inexpert status. The issues involved are too important for our self-understanding as human beings to be left to the disagreements of the experts alone and I think even a quizzical theoretical physicist should be allowed to have his say. There is a real sense in which we all have standing in this field, for we all know *from the inside* what it is like to live a mental life.

Examining the phenomena

Much of contemporary discussion presents us with an extraordinary situation. One might have supposed that the foundational phenomena from which that discussion must begin would be those experiences of intention to act, belief about truth or falsehood, experience of pain or pleasure, perception of colour or musical notes, which seem to be the raw material of the mental life. Indeed, herein appears to lie the very reason why we wish to introduce the category of mental events and distinguish them from the category of physical events. Pain is a very clear case in point. David Hodgson says

> One can readily distinguish the actual subjective feeling of pain from the physical processes in the brain which must be associated with that feeling, and from the pain reactions (withdrawal, grimacing, rubbing, crying, etc.) which may accompany feelings of pain and may be observed by others. The distinction between conscious mental events and physical events is probably clearer in the case of pain than most other examples.[1]

I agree, but there are philosophers of mind who do not.[2] They characterize such an understanding as 'folk psychology' and believe it to be as misleading about the nature of reality as is the common parlance that the sun 'rises'. Folk-chatter about mind is asserted to be just a manner of speaking about something else, such as information processing or the prediction of behaviour.

This sceptical view has been roundly criticized by John Searle. He points out that 'folk theories have to be in general true or we would not have survived'.[3] Folk physics may not have got the structure of the solar system right, but it was never in error about the undesirable consequences of walking off a high cliff. Searle lists three propositions:

1. In general, beliefs can be either true of false.
2. Sometimes people get hungry, and when they are hungry they often want to eat something.
3. Pains are often unpleasant. For this reason people often try to avoid them.[4]

and he goes on to comment concerning these conclusions of folk psychology that 'It is hard to imagine what kind of empirical evidence could refute these propositions'.[5] His verdict on many of his fellow philosophers of mind is that there is 'no other area of contemporary analytical philosophy where so much is said that is so implausible'.[6] The instinct of a scientist is to be a bottom-up thinker – to start in the basement of particularity and build upon the evidence of experience. No account of mind will be satisfactory that seeks to eliminate or devalue our basic mental life.

Thought experiments

At the same time that some philosophers of mind reject taking pain and perception seriously, there is a tendency to turn instead to seek a source of insight from bizarre thought experiments. The staple diet of such discussions is provided by the activities of Mad Scientists or Callous Brain Surgeons. The former maintain your brain in a vat and feed it with cunningly contrived stimuli,[7] or they construct a teletransporter that will dissolve your body here and reconstruct an exact replica somewhere else.[8] The surgeons divide brains into right and left hemispheres and transplant them separately into conveniently available other bodies.[9] One feels that these fearsome figures play the same sort of philosophically teasing role that the maliciously misleading demon, deceiving our minds, did for René Descartes. No doubt, reflection on these ideas can provoke some interesting discussion, but the insights they present often seem to be treated with a seriousness that would be appropriate only if we actually knew them to be realizable possibilities. Assuming that this is so is clearly tantamount to a prior decision about many of the aspects of the nature and interrelationship of mind and brain. Implicit in what is being said is a close degree of identification of the two. Someone like the philosopher of religion Richard Swinburne, who takes a dualist view that mind and brain are separate entities, is, from his position, right to state about split-brain transplants that 'however much we knew in such a situation about what happens to parts of a person's body, we would not know for certain what happens to the person'.[10] I shall want to adopt a more integrated view of human nature than Swinburne's, but I do not think that entails my having to believe that such replications and transplantations are available to human beings with just a little more technical know-how than we happen to possess today. Human embodiment may be too delicate and fragile for such rough procedures to be feasible, and that very delicacy may be constitutive of the possibility of a mental life.

I would draw a clear distinction between these totally speculative thought experiments and the mental exercises used by Einstein and Bohr to explore the implications of relativity theory and quantum mechanics. The latter employed accepted physical principles and sought to follow through their consequences for particular processes such as measurement. In the case of the Mad Scientists, it is the principles themselves that are the matters of dispute.

Subjectivity

Central to mental life is the phenomenon of consciousness. The existence of human self-awareness is a basic fact about the universe in which we live and no account of that world will be a candidate for a satisfactory metaphysics which fails to say with Thomas Nagel that consciousness 'must occupy as fundamental a place in any credible world view as matter, energy, space, time and numbers'.[11] Attack on the problem is exceptionally difficult because of the self-relating character of consciousness and because its very existence is the foundation of the possibility of our having *knowledge* of anything. Searle comments on these problems.

> Indeed it is the very subjectivity of consciousness that makes it invisible in a crucial way. *If we try to draw a picture of someone else's consciousness, we just end up drawing the other person* (perhaps with a balloon growing out of his or her head). *If we try to draw our own consciousness, we end up drawing whatever it is we are conscious of.* If consciousness is the rock-bottom epistemic basis for getting at reality, we cannot get at the reality of consciousness in that way.[12]

Science's strategy has been to regard the world and its contents as objects 'out there', available for our manipulation and interrogation without disturbing the mind's eye of the observer. The advent of quantum theory has only partially modified this objectivistic stance, for, in the most widely held interpretation of its meaning, it is the intervention of (impersonal) measuring apparatus that determines what the result of an observation will be.[13] The observing mind retains a degree of detachment, since the aim of the interpretative exercise is to secure an account consistent with intersubjective agreement.

What is an effective and particular investigative strategy for science, would become a metaphysical disaster if it were made into a rule for all. Yet, since the Enlightenment, that has been the attitude adopted by almost all Western thinkers. Searle says that 'Since the seventeenth century, educated people in the West have come to accept an absolutely basic metaphysical presupposition: *Reality is objective*'.[14] This is fatal for a philosophy of mind because, as Searle goes on to note, 'the ontology of the mental is essentially a first-person ontology'.[15] Pain is always *somebody's* pain. That is why it can never properly be encapsulated in objective talk about patterns of firing neurons. There is an inescapable personal privacy about the mental, hence all the puzzles about whether you mean by 'blue' what I mean by 'blue'. We can, of course, pick the same one out of a set of coloured discs, but do we have the same perception of its appearance?

We have to acknowledge that we view reality from the particular perspective of our individual experience. Failure to do so would be to deny the basis of all actual knowledge. Consciousness is not the epiphenomenal garnishing of a fundamentally objective and material reality; it is the route of our access to all reality. Refusal to take it seriously subverts the whole metaphysical enterprise. Searle is right to say that 'More than anything else, it is the neglect of consciousness that accounts for so much barrenness and sterility in psychology, the philosophy of mind, and cognitive science'.[16]

The adequate recognition of irreducible subjectivity does not, however, dissolve the discussion into accounts of a myriad of private worlds. Nor are we condemned to solipsism. Not only are there the scientific agreements we are able to reach about the nature of the physical world, but there is also a degree of mutually comprehensible insight into human experience that permits the existence of literature (with its explorations of a personal world in a way recognizable by a variety of readers) and the creativity of the arts (which invoke a shared encounter with the beautiful). Any theory of consciousness will have to take into account that our individual perceptions are capable of at least the partial degree of reconciliation that makes sense of our intuition that other minds exist and that we live in a common world.

Evolution

There are some who will see a mutuality of minds as being genetically imposed by the shared need to survive in the world as it is. It seems plausible to say that, if our thoughts did not conform to the realities of everyday physical experience, then we should not have continued for long to be successful in the struggle for existence. But that remark conceals an unsolved puzzle about the relationship between consciousness and evolution. It is clear that survival requires effective interaction with the environment, but it is far from clear that this also demands the quality of self-awareness. Hodgson comments 'Information processing in accordance with effective procedures or formal systems does not require consciousness'.[17] In fact, with its necessary focus of attention and the consequent possibility of diversion from the apprehension of danger, one might even consider that consciousness had certain negative consequences for survival. An efficient robot might be better endowed to deal with hazards than a thinking person. Safety warning systems in chemical production plants exploit this fact.

Dualism and monism

There is, however, a very important further question about the nature of consciousness, which arises from the recognition that we live in a universe which throughout its fifteen billion years history has been an evolutionary world. Continuously developing process appears to link a hot energetic quark soup (the universe when it was 10^{-10} seconds old) with the home of saints and scientists today. The emergence of consciousness seems to me to be far the most striking and significant development in all that long cosmic history, but it seems natural to seek to understand it as the full flowering of a potentiality always present, rather than the injection from outside (even by a benevolent Creator) of a totally new and distinct kind of substance. This thought, together with evidence for human psychosomatic unity provided by the effects of drugs and brain damage and the phenomenon of occasional remissions of grave illness in those with a certain set of the mind, does not encourage a Cartesian dualist view of reality. That kind of metaphysics always had a problem in explaining how extended matter and thinking mind interacted with each other; how my mental intention of raising my arm is translated into the physical action of my doing so. I think those difficulties have now become so intense that twentieth century thinking about the mind/brain problem must seek a solution along more integrated lines. The philosophers call the search for this kind of understanding 'dual-aspect monism'. The world is made of one sort of 'stuff', but of a subtlety that reduces it neither to mere matter nor to pure mind. Our encounters with the material and the mental are contrasting poles of a single experience. The material and the mental are to be given equal force in assessing the adequacy of our metaphysical conjectures. How such a dual-aspect monism could actually preserve the true character of the mental without an unacceptable reduction to a mere epiphenomenal froth on the surface of the material is, of course, an extremely difficult question to answer. I shall make some tentative suggestions toward the end of this chapter, but they will necessarily be speculative and inadequate. I think we face here a problem that will, on the most optimistic forecast, require centuries of effort to produce real progress towards its solution.

One of the attractions of a dualist view appeared to be that the notion of the soul as a detachable spiritual entity, associated with the body but not identified with it, provided a means of understanding the basic human intuition that each of us is a continuing self, the abiding subject of our experience. That little boy with the shock of black hair that I see in the fading photograph was *me*, who am now a balding academic in later middle age. The child who was so good at arithmetic but had some trouble in learning to read was *me*, who am now a scientist given to writing about science. Externally, that seems testified to by the (in principle, traceable) physical history linking that youngster with the current President of Queens'. Internally, it is testified to by its being my present memory that recalls those early school successes and problems. If we no longer have a spiritual soul to act as the invisible source of that intuition of selfhood and to be the carrier of its continuity, what shall we put in its place?

A place for the soul

I do not think we have to abandon all talk of the soul but we shall have to redefine it to accord with what reality is actually like. My soul is the *real me*, but that is neither a spiritual entity temporarily housed in the physical husk of my body, nor is it just the matter that makes up that body. After all, the latter is continuously being changed through the effects of eating and drinking, wear and tear. We have very few atoms in our bodies today that were there a few years ago. If there is a bodily basis for the experience of the continuity of the self, it lies in the almost infinitely complex information-bearing pattern in which those atoms are organized. This 'pattern' – I am using the word in some highly generalized sense that I do not know how to define properly – is continuously transformed (we acquire new memories, for instance) but the very continuity of that change is the basis for our continuing selfhood.

The view of the soul that I have been trying to express would not have surprised St Thomas Aquinas. He took from Aristotle the idea that the soul is the 'form' (pattern) of the body. Aquinas wrote that 'Therefore the soul, as the primary principle of life, is not a body but that which actuates a body'[18]. He explicitly thought that a Platonic separation of soul and body did not fit the facts of human existence.

Personal identity

Among those facts, I take it, there is our self-understanding that we are continuing persons. I cannot regard the self, as the philosopher Daniel Dennett does, as being a useful fiction spun out of the web of mental events. He likes to refer to the self as being 'a Centre of Narrative Gravity'[19], a useful convention like the physical centre of gravity of a body, but not an identifiable existent. I do not regard our notion of selfhood as being just a convenient manner of speaking (By whom? one might ask).

Another philosopher, Derek Parfit, has given a careful discussion of these problems and his conclusion is that what matters is not personal identity, which he thinks to be an elusive notion, but the sort of psychological continuity that corresponds to our impressions of a memory of our past. In his view

> *What matters is Relation R:* psychological connectedness and/or continuity, with the right kind of cause.
> Since it is more controversial, I add, as a separate claim
> In an account of what matters, the right kind of cause could be any cause.[20]

Parfit is led to his perplexities about personal identity because of the conundrums posed by alleged possibilities of brain division and transplantation and the production of replicas of people. If the Mad Scientist feeds me into his duplicating machine, the outcome is two people who are R-related to me-as-I-was, but Parfit regards as being empty the question of which is then me. That is why he believes that it is not 'personal identity' that matters. (Let us suppose that the intricacy of the replicating machine is such that one cannot attempt to settle the issue by an argument of physical continuity. Perhaps I am broken up in the course of analysing the pattern of my embodiment and then two copies of that pattern are constructed from new material.) If I am told that one of these replicas will be instantly slain and the other promoted to wealth and power, Parfit believes that I have no legitimate interest now in which will be the sufferer and which the gainer in that future event.

Well, philosophy is wonderful, but peculiar premises may lead to peculiar conclusions. Maybe these fearsome thought experiments are more a means for metaphysical legerdemain than a reliable guide to the nature of reality. In fact, I am inclined to accept something like Relation R as a possible way of thinking about *continuing personal identity*, but with the 'right kind of cause' more carefully construed. If 'any cause' is intended to indicate that actual technological development will put this possibility within the future grasp of humanity, I am disposed to reject this promethean claim. I suspect that the power of producing 'psychological connectedness and/or continuity' is divine in character, rather than human. (Its possibility for God has been the basis of my own understanding of the coherence of the Christian hope of a destiny beyond death.[21]) Such a power is safely possessed by the faithful Creator, who can be relied upon to resist the magical temptation to generate doppelgangers, which Mad Scientists apparently find so seductive. It is also possible that Relation R requires further elucidation as the sign of continuing personal identity. In a vague but suggestive way, one could suppose that the complex information-bearing pattern that is my soul could be characterized, through all its change, by certain persistent characteristics of structure (a mathematician would say, certain 'invariants') which make it uniquely me and no other person.[22]

Reductionism

Of course, the simplest kind of monism would be of a single-aspect kind – there is one sort of stuff whose nature is already known to us. The bluntest form of this assertion is made by the physical reductionists, for whom mind is a mere epiphenomenon of matter. The implausibility of this approach should by now be clear. Its essence is to deny the reality and specificity of the mental. It abolishes the problem by neglecting the evidence of experience. Nagel comments trenchantly that 'To insist on trying to explain mind in terms of concepts and theories that have been devised exclusively to explain nonmental phenomena is, in view of the radically distinguishing characteristics of the mental, both intellectually backward and scientifically suicidal.'[23] It is also destructive of the very rational argument it seeks to use. Long ago J. B. S. Haldane said that 'if materialism is true, it seems to me that we cannot know it is true. If my opinions are the result of chemical processes going on in my brain, they are determined by chemistry, not the laws of logic'.[24] Haldane later retracted this argument, as he was impressed by the logical operations performed by the mechanical hardware of a computer. I will discuss computational theories of mind very shortly and thus explain why I think the retraction was mistaken. For the moment let me just say that a successful programme requires a programmer, and where could such an entity be found in the impersonal world of materialism?

Some will cast evolution for the role of the Great Programmer. Certainly survival strategies will have some genetically based consequences for human behaviour, but that seems to amount to much less than what we are trying to understand. One can think of genes operating in the context of the living cell as providing the pattern for an adaptively structured body whose perceptual and reactive capabilities will have been constrained by the operation of natural selection. However, what can be understood in this way falls far short of the actual subtlety and fruitfulness of the human mind.[25] Our scientific, aesthetic, moral and spiritual powers greatly exceed what can convincingly be claimed to be needed in the struggle for survival, and to regard them as merely a fortunate but fortuitous by-product of that struggle is not to treat the mystery of their existence with adequate seriousness. This fact of our powers of mind encourages in me the thought that reality includes a noetic realm of the mental in which we participate and in which we have interacted in parallel with our evolution in the realm of the physical.[26] Dual-aspect reality implies dual-aspect development. My scepticism about the adequacy of a purely Darwinian, single-aspect, account is not removed by the interesting results concerning the ability of neural networks to 'learn' by adjusting the weightings of internal connections.[27] Such activity still requires for its execution that there be a 'learning algorithm', programmed in as the controller acting in response to inputs specifying the degree of error.

Materialists are given to drawing blank cheques on unknown intellectual accounts in order to rescue themselves from their implausible elimination of the mental. One such strategy is what Searle calls 'the "heroic-age-of-science" manoeuvre'.[28] Physical science has explained so much, why should it not eventually explain everything? Two comments can be made. One can be expressed in a favourite phrase of the sharp-tongued theoretical physicist, Wolfgang Pauli: 'No credits for the future'. Pauli liked to use this to rebuke those who waved their hands and expressed the vague hope that in the end it would work out all right. The other comment is to note that, even within its own domain, the explanatory power of physics has resulted from that subject's ability radically to revise its account of the physical world in the light of increasing experience of it. It would have been impossible to understand superconductivity without the revolutionary discoveries of quantum theory, which so substantially modified the Newtonian account of what matter is like. Consciousness is surely a much more profound phenomenon than superconductivity and its understanding may be expected to call for correspondingly much more radical revision of contemporary thought.

However, in the present state of ignorance about the nature of the mental and its relationship to the material, it is true that no one can get away without some considerable degree of handwaving. While Searle opposes a physicalist reductionism, his own preferred theory, that mind is an emergent property of matter in complex organization, is certainly not without its own difficulties. He writes that

> Consciousness is a higher-level or emergent property of the brain in the utterly harmless sense of 'higher level' or 'emergent' in which solidity is a higher-level emergent property of H_2O molecules when they are in a lattice structure (ice), and liquidity is similarly a higher-level emergent property of H_2O molecules when they are, roughly speaking, rolling around on each other (water).[29]

A little reflection leads to the recognition that the claimed analogy is unconvincing. Solidity or fluidity are energetic properties of aggregations of matter and it is scarcely surprising that they can emerge from energetic transactions among constituent molecules. But mental events do not have the character of energetic processes. Rather they are concerned with something that has the character of perception or intention or thought. Searle completely neglects the yawning gap between patterns of neuron firings and perceiving a patch of pink. Dennett makes the same mistake when he says 'But why should consciousness be the only thing that can't be explained? Solids and liquids and gases can be explained in terms of things which aren't themselves solids or liquids or gases'.[30] In a broad sense, like can only explain like. Energetic properties are explained by energetic transactions. The mental requires more than that because it is incommensurable with the merely material. In a crude dichotomy between software and hardware, it has more the character of 'information' than of energy.

The computer analogy

Perhaps in the end of the previous section we can find a clue. Functionalists suggest that a theory of mind should be a theory of information processing. They urge us to set aside the perplexities of conscious introspection and regard the question as one of linking input and output through the 'black box' processor of the mind/brain. What the brain *does* is what counts for a functionalist. A particularly popular and somewhat more explicit version of this view is one that uses a computer model for what the brain is doing. Once again, however, the essence of the mental has been abandoned in an attempt to gain a quick solution to the problem of its relation to the material. Roger Penrose has given arguments from mathematics which he believes show that human thought transcends the execution of computational algorithms.[31] He draws our attention to the proof of Gödel's theorem, which depends upon exhibiting a proposition that we can 'see' to be true but which is neither provable nor disprovable within the logic of a closed system. Penrose takes a Platonic view of mathematics, whose discoveries are held to result from the exploration of a noetic world of a kind somewhat similar to that which I have suggested is a part of reality.

Another line of argument leading to the same conclusion of the inadequacy of computer modelling depends upon taking seriously what Hodgson calls 'plausible reasoning'[32], those intuitive powers that again cannot be reduced to algorithmic procedures. In the well-known words of Michael Polanyi 'we know more than we can tell'[33], a property a computer could not possess, for its logical workings are laid bare.

The strongest argument against computer functionalism lies in drawing the distinction between syntax and semantics, between logical operations and meanings. One of the most celebrated ways of making this point is through Searle's parable of the Chinese room.[34] You are sitting in a closed office. Slips of paper marked with squiggles are passed to you from outside. A big book tells you that if you receive such and such a set of squiggles you are to hand out in reply such and such another set of squiggles. It turns out that the squiggles handed in are questions in Chinese and the squiggles you hand out, in accordance with the instructions of the rule book, are the appropriate answers in Chinese. You are acting as a computer whose (syntactical) operations are perfectly in accord with Chinese information processing but, nevertheless, you have no (semantic) understanding of the Chinese language. Regard yourself as an analogue of the computer brain and you will see that such a model of the mind completely fails to describe the access to meaning that is basic to human thought.

If there were 'understanding' located in the Chinese room, it would be in the big book (the programme) and not with you (the computer). More accurately, understanding is outside the room in the one who compiled the book (the programmer). One of the fallacies in computer modelling is to neglect the role of the programmer. The theoretical chemist Giuseppe Del Re rightly says that 'it is difficult to conceive of the self as a super-programme and not as a programmer'.[35]

A very sophisticated attempt to salvage some form of computer model has been made by Dennett in his ambitiously entitled book, *Consciousness Explained*. He emphasizes the need to think of the brain, not as a classical computer of the kind described by John von Neumann, but as a flexible and somewhat anarchic collection of parallel processors. (The need to adopt a parallel processor picture of the brain is endorsed by the recognition that neural response times require something quicker than the von Neumann machine model could provide.) He calls his idea the 'Multiple Drafts model', according to which 'all varieties of perception – indeed, all varieties of thought or mental activity – are accomplished in the brain by parallel, multitrack processes of interpretation and elaboration of sensory inputs. Information entering the nervous system is under continuous "editorial revision" '.[36] One of the objects of his scorn is what he calls 'the Cartesian Theater', the notion that there is some internal screen scanned by the 'homunculus' who is 'the Central Meaner'. One can see the unsatisfactoriness of the hypothesis of the little man in the middle, but the problem is what to put in its place to represent our actual experiences of conscious thought and perception. The answer we are given is that, somehow, the Darwinian struggle between the multiple drafts of the parallel processors produces a sequence of particular offerings that win out above the parallel hubbub to create the appearance of a single-track 'von Neumann bottleneck'. Dennett concludes that 'Human consciousness . . . can best be understood as the operation of a *"von Neumanesque"* virtual machine *implemented* in the *parallel architecture* of a brain that was not designed for any such activities'.[37] This happy evolutionary accident is dubbed 'the Joycean machine', to encourage the feeling that it has something to do with the human stream of consciousness.

Despite the ambitious title, it seems to me that Dennett's book nowhere adequately addresses the issue of self-awareness. No contact has actually been made between computer-talk and mental experience. The yawning gap remains unbridged. It is not clear that the virtual von Neumann machine represents any advance on the Cartesian theatre. It simply sits on one side of the gap (computation), as unsatisfactory in its way as the idea of an homunculus sitting on the other side of the gap (awareness).

I am not unsympathetic to trying to use computer analogies to get some extremely modest and primitive insight into the problems under discussion. What I am opposed to is the claim that what in fact is only a preliminary exercise of very limited scope is the total solution of the problem. People such as Dennett seem to suppose that all one needs to do is to add a few simple ideas, like the Joycean machine, to standard computational theory and all is solved. That seems to me like saying in 1900 that all one had to do to cope with the problems of atomic physics was to add Max Planck's notion of packets of energy to Newtonian mechanics and all was solved. Insightful though Planck's discovery was, the true explanation of atomic physics required a revolutionary transformation of our ideas of the nature of the physical world, brought about by quantum theory. It would surely be surprising if the comprehension of consciousness did not call for at least an equal upheaval in our understanding of reality.

One might make similar comments about Francis Crick's approach to consciousness through the deliverances of neuroscience. His so-called 'Astonishing Hypothesis' is that "You", your joys and your sorrows, your memories and your ambitions, your sense of personal identity and free will, are in fact no more than the behaviour of a vast assembly of nerve cells and their associated molecules'[38]. The bulk of his book is concerned with a detailed and interesting account of what is known about parallel pathways in the brain that process visual information. At the higher levels of brain activity, this information must be 'bound' together to construct representation of reality, in ways not currently understood. This funnelling of parallel processing into serial integration is described tendentiously as an ' "attentional" mechanism', with the suggestion that this could be 'the "neural correlate" of consciousness'.[39] The quotation marks are indeed well deserved, for a quite illicit slide is taking place from computer-like model construction to conscious awareness. No discussion of neuronal firing rates, or correlated neuron activity, will bridge the yawning gap between current neuroscience and the experience of perception. In his conclusion, Crick confesses that he believes 'the correct way to conceptualize consciousness has not yet been discovered and that we are merely groping our way towards it'[40]. I believe that this exploration would have been better conducted if Crick had not tethered himself to the ideological stake of the physically reductionistic 'Astonishing Hypothesis'.

Some 'pre-Socratic flailing about'

Nagel says 'Too much time is wasted because of the assumption that methods already in existence will solve problems for which they were not designed'.[41] In a splendid phrase, he tells us that to attempt the theory of a dual-aspect monism today, is 'probably nothing more than pre-Socratic flailing about'.[42] Of course, people like Thales and Anaximenes were two and a half millennia too early to solve the problem of the nature of matter, but it was a worthwhile venture for them to start thinking about the possibility that the variety of the world is the result of the different states of a small number of basic kinds of stuff. In a similar way, I think we should not refrain altogether from waving our hands in a dual-aspect direction, even if we have to be very modest and tentative about what we might hope to achieve thereby.

The heart of the dilemma is, on the one hand, the apparent disjunction between the mental and the material and, on the other, their intimate connection in our psychosomatic experience. In such perplexity an analogy is like a plank to a drowning man. Physics provides us with examples where apparently contradictory properties can be found to be possessed by the same entity in appropriately different circumstances. A prime example is the wave/particle duality of light.

A particle is a small bullet-like entity; a wave is a spread-out flappy kind of thing. In the common sense logic of everyday thinking they are about as distinct from each other as it is possible to be. Yet, light exhibits both kinds of behaviour, according to the mode in which it is investigated. Ask it a particle-like question (photoproduction, the ejection of electrons from metals) and it gives a particle-like answer. Ask it a wave-like question (the two-slits experiment, interference fringes) and it gives a wave-like answer. Since the discovery of quantum field theory, we have understood how such ambidexterity is possible.[43] Forbidden by the rigid clarity of a Newtonian ontology, it is allowed by quantum uncertainty. You could never build a wave out of finite collections of particles, but a wave-like state is one with an *indefinite* number of particles making it up. That is how the trick is done.

This combination of apparent opposites within the openness provided by intrinsic indeterminacy, is called *complementarity*. I have suggested that mind and matter might be complementary poles of the single stuff of a dual-aspect monism.[44] One must then go on to speculate what might be the source of the indefiniteness that would make this conceivable.

One obvious answer to consider would be quantum theory itself. Among those who have explored this approach are the philosopher Michael Lockwood[45] and the jurist David Hodgson[46]. The latter summarizes his position as being that

> Mind can to some extent be said to be a function of brain, but only if the brain is understood not as the detectable macroscopic object, but as the quantum reality underlying *both* this object *and* the mental events of consciousness. Mind and brain are two manifestations of, and viewpoints towards, a single reality; but with important differences, in particular in relation to the development over time of this reality and (specifically) the causes and explanations of such developments.[47]

The last remark is directed towards Hodgson's conjecture that the indeterminacies and non-localities of quantum events in the cortex are constitutive of the character of mental experience.

It is certainly possible that microscopic quantum effects play a part in the mystery of the relationship of mind and brain[48] but I am personally reluctant to suppose that this is where the whole solution is to be found. It is the macroscopic world that we perceive and in which we execute our intentions. The basic experience of the self seems to be that of the whole embodied person. I have been disposed, therefore, to look for a significant source of complementary indefiniteness in the processes of the macroscopic physical world. My proposal is that an interpretation of the dynamic theory of chaos might provide the necessary clue.[49] Here, I can only outline the argument.

Most of the macroscopic physical world is composed of systems so exquisitely sensitive that their behaviour is vulnerable to radical change by the slightest disturbance in their circumstances. In consequence, such systems are intrinsically unpredictable and they can never be treated in isolation from their environment. Their behaviour is not, however, completely haphazard, for it is found to be confined within a certain range of possibilities given the name of a 'strange attractor'. Chaotic systems are both orderly and disorderly at the same time. Everyone agrees that these surprising results place severe epistemic limitations on the possibility of our knowledge of chaotic complex systems. My instinct as a physicist is not to regard this as just a tiresome principle of ignorance. Scientists are realists at heart and so they believe that what we can or cannot know is a reliable guide to what is actually the case. This conviction prompts the construction of a metaphysical conjecture which regards the unpredictability of chaotic systems as an indication that in reality they possess some degree of ontological openness in their behaviour. It also treats the deterministic Newtonian equations from which the theory apparently arose as being *approximations* to reality, valid only in those unusual circumstances in which the systems can truly be treated as isolable components of the physical world. I have called this view *contextualism*: the behaviour of parts is not independent of the context of the whole.

This proposal runs counter to conventional wisdom in relation to chaos theory. The latter supposes that the principal lesson to be learnt is that complex and apparently random behaviour can originate from an underlying deterministic simplicity. This is certainly a mathematical fact worth knowing. When one comes to apply the theory to the physical world, however, it seems to me that a realist strategy encourages a different understanding. Aligning epistemology and ontology as closely as possible to each other favours an open rather than a deterministic interpretation of the unpredictabilities all acknowledge to be present. I have sought to discuss this contentious issue in greater detail in my other writings.[50]

The ontological openness referred to is construed as implying that the causal principles that actually bring about the future state are not only the energetic exchanges between constituents but there is also scope for the operation of holistic causal principles. Consideration of how the latter might operate in bringing about the particular way in which a chaotic system traverses the labyrinth of possibility contained within the fixed energy of its strange attractor suggests that these holistic causal principles would have the form of pattern-creating agencies. They could be thought of as 'active information'. At work in the world, then, are both the bottom-up causality of physics and the top-down causality of this new kind. Such complementary causality, by local energy exchanges and by holistic active information, seems like the *glimmer* – I claim no more than that – of a way of thinking which would accommodate a duality of mind–matter. That is my particular way of pre-Socratic flailing around. In its complementary way it does nothing to deny the insights of physics. At the material pole of reality, if you split me apart into my constituents, you will just find quarks and gluons and electrons. Yet you will also have destroyed me. The self resides at the other, holistic, pole of reality. That explains its elusiveness in the reductive analyses of materialism or computer functionalism. In their different ways, both the Cartesian theatre and the Joycean machine are as misguided as would be attempts to make a wave out of finite collections of particles. The infinite regressions, threatened in such analyses, are signs that one is looking in the wrong place, scrabbling around among the pieces for what can only be found in the whole.

An account of reality without a proper account of mind would be pitifully inadequate. The difficulty of accomplishing much in our present state of knowledge should not cause us to abandon the task or to trivialize it into some facile and unconvincing form of reductionism. We have to be realistic enough, and humble enough, to recognize that much of what is needed for eventual understanding is beyond our present grasp. With Nagel, I believe that when such comprehension comes 'it will alter our understanding of the universe as radically as anything has to date'.[51]

NOTES

1. D. Hodgson, *The Mind Matters* (Oxford University Press, 1991), p. 40.
2. E.g. eliminative materialists such as P. M. Churchland, *Matter and Consciousness* (MIT Press, 1984).
3. J. Searle, *The Rediscovery of the Mind* (MIT Press, 1992), p.59.
4. *Ibid.*, p. 62.
5. *Ibid.*
6. *Ibid.*, p. 3.
7. D. Dennett, *Consciousness Explained* (Little, Brown & Co., 1991).
8. D. Parfit, *Reasons and Persons* (Oxford University Press, 1984).
9. *Ibid.*
10. R. Swinburne, *The Evolution of the Soul* (Oxford University Press, 1986), pp. 148–9.
11. T. Nagel, *The View from Nowhere* (Oxford University Press, 1986), p. 8.
12. Searle, *Rediscovery*, p. 96.
13. See J. C. Polkinghorne, *The Quantum World* (Longman, 1984), ch. 6.
14. Searle, *Rediscovery*, p. 16.
15. *Ibid.*, p. 20.
16. *Ibid.*, p. 227.
17. Hodgson, *Mind*, p. 157; a similar criticism is made by Nagel, *View*, pp. 78–82.
18. Quoted in B. Davies, *The Thought of Thomas Aquinas* (Oxford University Press, 1992), p. 212.
19. Dennett, *Consciousness*, passim.
20. Parfit, *Reasons*, p. 215.
21. J.C. Polkinghorne, *Science and Christian Belief* (SPCK, 1994), pp. 163–4.
22. Cf. J. C. Polkinghorne, *One World* (SPCK, 1986), p. 85.
23. Nagel, *View*, p. 52.
24. Quoted in Hodgson, *Mind*, p. 150.
25. See the discussion of Nagel, *View*, pp. 78–82.
26. I am indebted to Professor Edward Oakes SJ for an interesting conversation on this point. See also J. C. Polkinghorne, *Science and Creation* (SPCK, 1988), ch. 5.
27. See F. Crick, *The Astonishing Hypothesis* (Simon and Schuster, 1994), ch. 13.

28. Searle, *Rediscovery*, p. 5.
29. *Ibid.*, p. 14.
30. Dennett, *Consciousness*, p. 455.
31. R. Penrose, *The Emperor's New Mind* (Oxford University Press, 1989), ch. 4. See also *Shadows of the Mind* (Oxford University Press, 1994).
32. Hodgson, *Mind*, ch. 5.
33. M. Polanyi, *Personal Knowledge* (Routledge and Kegan Paul, 1958), passim.
34. J. Searle, *Minds, Brains and Science* (BBC Publications, 1984), ch. 2.
35. G. Del Re (ed.), *Brain Research and the Mind–Body Problem* (Pontifical Academy of Science, 1992), p. 26.
36. Dennett, *Consciousness*, p. 111.
37. *Ibid.*, p. 210.
38. Crick, *Hypothesis*, p. 3.
39. *Ibid.*, pp. 203, 207.
40. *Ibid.*, p. 255.
41. Nagel, *View*, p. 10.
42. *Ibid.*, p. 30.
43. See J. C. Polkinghorne, *The Particle Play* (W. H. Freeman, 1979), ch. 5.
44. Polkinghorne, *Creation*, ch. 5.
45. M. Lockwood, *Minds, Brains and the Quantum* (Blackwell, 1989).
46. Hodgson, *Mind*.
47. *Ibid.*, p. 381.
48. This view is also taken by Penrose (ref. 31), though he believes that a new physics is also needed that will lead to simultaneous resolution of problems in quantum measurement, gravity and consciousness.
49. Polkinghorne, *Creation*, ch. 3; *Science and Providence* (SPCK, 1989), ch. 2; *Reason and Reality* (SPCK, 1991), ch. 3; *Belief*, ch. 1.
50. Polkinghorne, Creation, ch. 3; *Science*, ch. 2; *Reason*, ch. 3; *Belief*, ch. 1.
51. Nagel, *View*, p. 51.

6

What does it mean?

Blind chance or purposeful Creator

One of the consequences of a religious view of the world (to which I subscribe) is the claim that there is a meaning to life and a purpose being fulfilled through the unfolding of history. There was a time when the discoveries of science appeared to give support to such an assessment. Newton's conclusion that the great variety of motions present in the solar system are the consequences of a single simple law of universal gravitation seemed to him to be an insight into a marvellous design. In the General Scholium attached to the second edition of the *Principia* he wrote,

> This most beautiful system of the sun, planets, and comets, could only proceed from the counsel and dominion of an intelligent and powerful Being ... This Being governs all things, not as the soul of the world, but as Lord over all ...[1]

If physics discerned a wonderful piece of cosmic clockwork, biology had even greater marvels to tell as it considered the aptness of living forms for functioning in their environment. In 1691, the distinguished Cambridge naturalist John Ray published a popular and influential work entitled *The Wisdom of God in the Works of Creation*. A hundred years later this line of argument found its most famous exposition in William Paley's *Natural Theology*. The celebrated inference of the existence of a watchmaker from the contrived and purposive complexity of a watch was used as an analogy for the corresponding presumption that the elaborate perfection of living organism should lead to the conclusion that the mind and purpose of the Creator was to be discerned in the design of creation. The argument was carried further in the subsequent sequence of Bridgewater treatises, concerned with demonstrating the Power, Wisdom and Goodness of God as manifested in the Creation; illustrating such work by all reasonable arguments, as for instance the variety and formation of God's creatures in the animal, vegetable and mineral kingdoms; the effect of digestion and thereby of conversion; the construction of the hand of man and an infinite variety of other arguments; as also by discoveries ancient and modern, in arts, sciences and the whole extent of literature.[2]

David Hume had produced trenchant criticisms of such a natural theology, drawing attention to the apparent imperfections of nature (of which Ray was certainly conscious) and the anthropomorphic character of the discussion. Yet the argument from design was widely felt to be forceful until Darwin, in the *Origin of Species*, showed another manner in which the trick could be done. The patient accumulation of small differences, and their sifting through natural selection, provided a way in which an organism could attain the form necessary for effective survival in an environment, without having to invoke the Hand of a Designer as the direct source of the aptness achieved. Life had not sprung into being ready made, at peremptory divine command. It had developed slowly through a contingent process of trial and error, change and selection.

Certainly it was no longer possible to think of the complexity and fruitfulness of the living world as being the consequence of the execution of an inexorable divine blueprint in which it had been decreed from all eternity what should be the length of the giraffe's neck or what were the details of the optical system of the human eye. Something more flexible and innovative must now be considered as the story of life's coming to be and of its development. Yet it is not clear that the recognition of the role of a degree of historical contingency necessarily implies the complete denial of the discernment of purpose in what has been going on, though many modern biologists have wanted to impose that interpretation. Richard Dawkins wrote,

Natural selection, the blind unconscious automatic process which Darwin discovered, and which we know is the explanation for the existence and apparent purposeful form of all life, has no purpose in mind. It has no mind and no mind's eye. It does not plan for the future. It has no vision, no foresight, no sight at all. If it can be said to play the role of watchmaker in nature, it is the *blind* watchmaker.[3]

Yet, from the start, the matter has been seen to be more open to question than that. Charles Kingsley, an Anglican clergyman and contemporary of Darwin, welcomed the insights of evolution – for the popular myth of progressive science versus obscurantist Christianity is an historically ignorant account of the evaluation of Darwin's great discovery. Kingsley said that in the light of evolution scientists found that 'they have got rid of an interfering God . . . [and] have to choose between the absolute empire of accident, and a living, immanent ever-working God'.[4] He saw that God had not just produced a ready-made world; he had done something cleverer than that in making a world that could make itself. A similar thought was expressed by his contemporary Aubrey Moore. Writing in *Lux Mundi*, he asserted that Darwinism, by making untenable the picture of a deistic intervening God, had 'under the guise of a foe [done] the work of a friend'.[5] Instant creation had been replaced by the concept of a continuing creation. This idea still plays an important role in religious reflection on an evolutionary universe, finding a variety of expressions, as in the writings of Teilhard de Chardin[6] and Arthur Peacocke.[7]

Those who on the contrary wish to assert the 'blindness' of evolution[8], concentrate their attention on the undoubted contingency present in the process, as if that were the only aspect involved. They stress that mutations occur in ways that are not directly linked to survival needs. Yet these offerings of chance are selected by interaction with environmental necessity, in a process of remarkable fertility. It is possible to understand the shuffling explorations of possibility, which we choose to call 'chance', as being a means for realizing, in contingent ways, something of the rich potentiality present in the necessity of natural laws. Evolution depends upon an interplay between chance *and* necessity and it is disingenuous not to consider the 'lawful' side of what is going on.

Cosmic fruitfulness

There is an astonishing drive towards fruitfulness present in the unfolding process of the world, which has turned a newly formed Earth into the home of self-conscious beings in little more than three billion years. That may seem a long time, but astonishing things have to happen. The human brain (evolved in only a few hundred thousand years from a much more primitive hominid brain[9]) is far and away the most complex physical system we have ever encountered in our exploration of the universe. It seems impossible in the state of current knowledge to extract from evolutionary geneticists even the crudest estimates of the timescales over which such neural complexity might be expected to evolve. Nevertheless a number of physical scientists (and not just those who might be thought to have a covert religious agenda) believe it likely that neo-Darwinism is only *part* of the explanation. They suspect that other, hitherto undiscovered, organizing principles may be at work, driving the development of complexity. The physicist Paul Davies comments,

> Strong organizing principles are invoked by those who find existing physical laws inadequate to explain the high degree of organizational potency found in nature and see this as evidence that matter and energy are somehow being guided or encouraged into progressively higher organizational levels by additional creative influences. Such principles may be prompted by the feeling that nature is unusually efficient at conquering its own second law of thermodynamics and bringing about organized complexity.[10]

Later he says

> The very fact that the universe *is* creative, and that the laws have permitted complex structures to emerge and develop to the point of consciousness – in other words that the universe has organized its own self-awareness – is for me powerful evidence that there is 'something going on' behind it all. The impression of design is overwhelming.[11]

I have considerable sympathy with the belief that the fruitfulness of cosmic and terrestrial history is such that it is reasonable to seek to supplement current received evolutionary ideas with the operation of possible teleological laws of nature (and, indeed, with the insights of a theology of nature seen as a creation).[12] I have even speculated about how there might be room for the operation of such teleological influences without denying our current knowl-

edge of natural process.[13] Darwinian ideas provide partial insight
into the developing history of a fruitful world but it is certainly
not known that they tell the whole story. Rather than pursue these
hypotheses once again, I prefer to turn to other, more clearly established, scientific insights which suggest that modern science is not
in fact inhospitable to a metaphysical discernment of meaning and
purpose lying behind cosmic history.

Beautiful equations

The first consideration is that science is possible at all only because
the physical world has proved to be remarkably rationally transparent to us. We are able to understand it to an astonishing degree.
Most of the time we take this for granted. Of course, if we could
not make sense of the *everyday* world, we would hardly have been
able to survive in the struggle for existence. If we did not learn how
to make generalizations such as 'deadly nightshade is poisonous' or
'jumping off a high cliff leads to disaster', we would not last long.
Yet our ability to understand the physical world immensely exceeds
anything that is required for the relatively banal purpose of survival. Think of the strange counter intuitive subatomic world of
quantum theory. If you know where an electron is, you cannot
know what it is doing; if you know what it is doing, you cannot
know where it is. That is Heisenberg's uncertainty principle in a
nutshell. The quantum world is totally unpicturable for us, but
it is not totally unintelligible. I cannot believe that our ability to
understand its strange character is a curious spin-off from our
ancestors having had to dodge sabre-toothed tigers.

That seems even clearer when we recognize that it is *mathematics*
which gives us the key to unlock the secrets of nature. Paul Dirac
spent his life in the search for beautiful equations. That is a concept
not all will find immediately accessible, but among those of us
who speak the language of mathematics, mathematical beauty is a
recognizable quality. It is hard to describe but easy to recognize –
like most other kinds of beauty. Its essence lies in a certain economy
and elegance that leads to the mathematical property of being
'deep'. Dirac once said that it is more important to have beauty

in your equations than to have them fit experiment. Of course, he did not mean by that that empirical adequacy was unnecessary. No physicist could believe that. But if your equations did not appear to fit experiment, there were various possible ways out of the difficulty. Nearly always one has to solve the equations in some sort of approximation and maybe you had not hit on the right way to do that. Or maybe the experiments were wrong – we have known that to happen. At any rate, there was some sort of chance of snatching success from apparent failure. But if your equations are ugly, then there is really no hope for you. Time and again we have found that it is equations with that indispensable character of mathematical beauty which describe the nature of the physical world.

If you stop to think about it, that is a very significant thing to have discovered. After all, mathematics arises from the free rational exploration of the human mind. Yet it seems that our minds are so finely tuned to the structure of the universe that they are capable of penetrating its deepest secrets. Mathematicians have a very modest way of speaking, but even they are inclined to acknowledge this as a 'non-trivial' fact about the world.

Einstein certainly saw it that way. He once said that the only incomprehensible thing about the universe is that it is comprehensible. In the rational beauty and rational transparency of the physical world we see the threads of a deep meaning woven into the empirical tapestry of science. Those physicists, like Stephen Hawking, who, in speaking of the mathematical order of the physical world like to refer to reading the Mind of God, are in my opinion speaking better than perhaps they know, though there remains much more to the divine mind than physics will ever disclose.

The Anthropic Principle

For a second consideration we should turn to the Anthropic Principle. In our scientific imaginations we can consider universes similar to ours but differing in some aspects of their physical fabric. One of the simplest variations to consider would be one in which the intrinsic strength of one of the forces of nature was different from the value it takes in our universe. For example, one could make the fine structure constant α (which measures the strength of electromagnetism) different from our value of about $1/137$. I would have guessed that this change would have no drastic effect on the history of that other world. If α were bigger, matter would be more dense (it is electromagnetism that holds bulk matter together), so the 'people' of that world would be chunkier. But I would have expected that the evolutionary history of that universe would have produced its own kind of life – not *Homo sapiens*, of course, but maybe little green men. I would have been mistaken! A universe of that kind would have had a boring and sterile history. Evolution by itself is not enough. You cannot, if you want to fulfil the role of Creator, simply bring into being more or less any old world and just wait a few billion years for something interesting to happen. Only a very particular, a very 'finely tuned', universe is capable of producing systems of the complexity and fruitfulness to make them comparable to *anthropoi*. The interplay of chance and necessity requires the necessity to have a very special form if anything worthy (by our standards) to be called 'life' is to emerge. It is this surprising conclusion that has been called the Anthropic Principle.

It is worth looking in some detail at why a fruitful universe is though to have to be so special in its physical constitution.[14] Many reasons can be given and I shall only attempt to indicate something of the range of considerations involved.

In the first instance one needs to have the right kind of physical laws. Nature must not be too rigid, or else there will not be scope for the kind of flexible change that is the engine of evolution. Equally, nature must not be too floppy, or else there will be no persistence in the novel forms of organization that come into being. Quantum mechanical laws provide just that basic opportunity for the interplay of chance and necessity that seems essential for fertile development.

Fine-tuning

Next, the intrinsic force strengths need to lie within very narrow limits. The most striking example of this is the cosmological constant. It corresponds to a term that logically can be present in the field equations of general relativity (the modern theory of gravity) but appears to be absent in our world to the extent that its value is zero to within one part in 10^{120}. Were the cosmological constant not virtually zero to this high degree of accuracy, that would make the evolution of life impossible, either by producing instantaneous cosmic collapse (if its sign were negative) or by inducing extremely rapid expansion and consequent dilution (if its sign were positive). This is the most stringent of all the anthropic requirements and it can result only from an exquisite degree of cancellation between two contributions that combine to produce the total effect.

Less exacting, but still tight, limits apply to the other forces of nature. Take electromagnetism. The nature of chemical bonding requires that it be not significantly weaker than it is, yet, if it were somewhat stronger, rates of chemical reactions would be appreciably slowed down and the evolution of life correspondingly retarded. There are many detailed properties of matter that depend on the electromagnetic force and have anthropic consequences. Crucial to the possibility of life in the waters of Earth is the remarkable property that ice is lighter than water, so that freezing takes place from the top downwards and not from the bottom upwards. In consequence, ice forms a skin on the surface, which easily melts at warmer temperatures and which protects the aquatic creatures living underneath while frost endures. A lake solidified from the bottom up would take a long time to melt and it could not be expected to sustain life.

Gravity must be strong enough to cause stars and galaxies to condense, but not so strong as to enforce a cosmic collapse. A particularly sensitive balance between gravity and electromagnetism controls the way in which stars burn (producing long-lived stable sources of energy, essential for the development of life). If electromagnetism were only slightly stronger than it is in relation to gravity, all stars would be red and probably too cold for supporting life; if electromagnetism were relatively slightly weaker, all stars would be blue, intensely hot, and they would live for only a few million years – far too short a time for an evolutionary history to develop on one of their planets.

There are two sorts of nuclear force at work in our universe: the strong nuclear force that holds nuclei together and the weak nuclear force that causes some of them to decay. The latter played an important role in early cosmic history. If it had been significantly stronger, hydrogen would have readily burnt to helium and only that element would have been left to constitute the galaxies and stars as they began condense. There would then have been no water and no hydrogen-burning stars, which, we have noted, alone give the stable long-lived sources of energy needed for the development of life. If, on the contrary, the weak force had been a little weaker there would then have been no hydrogen left over after those hectic first three minutes in which the whole universe was hot enough to be the arena of nuclear reactions. The survival of some hydrogen requires an excess of protons over neutrons, which is derived from the decay of neutrons into protons, and a feebler weak nuclear force would have made that too slow a process to be effective. The strong nuclear force also has its anthropic bounds: a little stronger and protons would bind to form the diproton (again, no hydrogen); a little weaker and the deuteron becomes unbound, with disastrous consequences for the nuclear processes that make stars burn.

The nuclear processes in the stellar furnaces do not only provide energy. They also make the heavier elements, essential for the chemistry of life. We are made from the ashes of dead stars! Both of the nuclear forces play finely tuned roles in the delicate story of nucleosynthesis. The strong force is such that there is an enhancement (a resonance) in just the right place to enable three helium nuclei to stick together and make carbon. Fortunately, there is not another such enhancement present in the process whereby a further helium nucleus can stick to the carbon to make oxygen. This has the result that some oxygen is made but some carbon is also left behind. (It would be anthropically disastrous to turn all the carbon into oxygen.)

The weak nuclear force plays an essential role in the way some stars explode as supernovae (thus scattering their previous nuclear products out into the environment, where they can become part of the chemical composition of second-generation planets) and at the same time make essential heavier elements (such as zinc and iodine) that cannot be created in stellar interiors.

The sequence of reactions involved in synthesizing the range of nuclei needed for life is extremely complex and delicately balanced. Its unravelling has been a major achievement of twentieth century physics. One of the people who played a leading role in its discovery was the astronomer Fred Hoyle. In fact he *predicted* the existence of the carbon resonance that makes that link in the chain possible, before it was known experimentally, simply in order to make the process viable. Hoyle was very impressed with the 'quirks' that occurred in just the right places to enable the sequence of element building to be completed. He wrote,

> I do not believe that any scientist who examined the evidence would fail to draw the conclusion that the laws of nuclear physics have been deliberately designed with regard to the consequences that they produce inside stars. If this is so, then my apparently random quirks have become part of a deep-laid scheme. If not, then we are back again at a monstrous sequence of accidents.[15]

A home for life

We can also look to the circumstances of the universe in which we live, as well as to its physical laws, to see that it is not any old world that can be the home of life. One of the cosmos's most striking features is its size. We live on a planet circling an undistinguished star, lying among the hundred thousand million stars of the Milky Way galaxy, which is itself a pretty run-of-the-mill specimen among the hundred thousand million galaxies of the observable universe. Yet, we should resist the temptation to be daunted by the thought of such immensity. Those trillions of stars have to be around if we are to be around also to think about them. In modern cosmology there is a direct correlation between how big a universe is and how long it has lasted. Only a universe as large as ours could have been around for the fifteen billion years it takes to evolve life – ten billion years for the first generation of stars to generate the elements that are the raw materials of life, and about a further five billion years to reap the benefit of that chemical harvest.

One more example of fruitful circumstance must suffice. In our universe the neutron is about 0.1% more massive than the proton. Another way of saying it is that the neutron – proton mass difference is about twice the mass of an electron. If the difference were greater, neutrons would decay into protons inside nuclei, which would then be blown apart by electromagnetism, so that hydrogen would be the *only* possible element. If the difference were a little less, free neutrons would not decay into protons (an essential process in the early universe, as we have seen, for it yields the necessary cosmic presence of hydrogen). Incidentally, the fact that the electron mass is so much less than that of protons and neutrons implies that the nucleus of an atom is little affected by the motion of its orbiting electrons, permitting molecules to have a stable shape and position and so permitting solids to exist.

One could continue multiplying anthropic conditions of this kind but enough has been said to indicate the character of the scientific insights involved. We should now go on to ask what one makes of it all?[16] Some reply that nothing can be learnt from one instance and we have only one universe accessible to our inquiry. Yet the whole nature of the argument has been that we can *imagine* universes that are similar to our own and that when we do so we find that only those extremely 'nearby' in their physical characteristics would be capable of a fruitful history. In the space of ontologically conceivable universes, we are surrounded by a large sterile patch (so to speak). Adopting a philosophical parable told by the philosopher John Leslie[17], if there is a single fly on a big blank wall, its being hit by a bullet surely calls for some sort of explanation. Either a marksman has been at work or many shots were fired, one of which by chance hit this isolated target. I will return to the detail of this parable in due course.

Inflation

Another suggestion that has been made is that maybe the remarkable coincidences which the evolution of life seems to require are in fact entailments of a some deeper physical theory. It is possible to give an example of this happening. One anthropic necessity is that the very early universe should be characterized by a very precise balance between the explosive effect of the Big Bang, driving matter apart, and the attractive pull of gravity, drawing matter together. If these were out of kilter with each other, either the universe would rapidly become too dilute for anything interesting to happen (if expansion dominated) or it would recollapse before anything interesting had time to happen (if gravity predominated). If one makes the simplest extrapolation back to the Planck time when the universe was 10^{-43} seconds old (the earliest time at which one can even pretend to say anything sensible with current knowledge) the balance between the two seems to need to be better than one part in 10^{60}. Paul Davies neatly interprets this degree of accuracy as corresponding to hitting a target an inch wide on the other side of the observable universe![18] When this condition was first recognized, it was thought that this was a very delicate balance that had to be built into the initial conditions of the universe. Now, most physicists believe that it would be achieved, almost whatever the literally initial circumstances, by a process called inflation, which is thought to have intervened when the universe was about 10^{-35} seconds old. It is supposed that there was then a kind of boiling of space which would have had as its consequence that it subsequently left the universe in a perfect balance between expansion and gravitational attraction.

Yet not every conceivable universe could have had an inflationary scenario as part of its history. That possibility itself requires that the laws of nature (now referring to the Grand Unified Theory thought to lie behind presently observed forces) would have had to take a particular form. One anthropic condition has been replaced by another. There is, it seems to me, bound to be something specifically necessary to provide the basis of fertility. The universe does not have *a priori* to be quantum mechanical, or gravitational, though both of these aspects of its nature play essential roles in determining the character of its evolution of life. Even if (as some speculate, but which does not seem all that likely to me) there is only one fundamental theory incorporating quantum gravity, it would still surely be remarkable that that unique possibility *also* provided all that was necessary for life's evolution; that is to say that gravity and quantum theory were not only anthropically necessary but also anthropically sufficient.

The carbon principle

Much the most difficult to evaluate of the criticisms of anthropic arguments is that which points out that the principle should really be called the Carbon Principle, or at most the Nuclear Principle, since so many of its conditions relate to the generation of the chemical elements necessary for carbon-based life, together with the resulting properties of matter. Isn't this just a lack of imagination? Might not other universes have their own forms of 'life', totally different from any that we can conceive but perfectly appropriate to their physical circumstances?

Something like consciousness seems to demand an immensely complicated physical carrier. There are as many neurons in our brains as there are stars in the Milky Way (10^{11}) and their interconnections are fantastically complex. It is impossible to say what radically different ways there might be for generating comparable complexity in totally different circumstances, but those who rely on this possibility to dismiss any significance in the fine-tuning of our own universe are drawing intellectual blank cheques on totally unknown accounts. I conclude that it is reasonable to continue to discuss the question of what significance attaches to these scientific insights.

A minimal response would be represented by what some call the Weak Anthropic Principle: the existence of human life imposes certain conditions on the universe and we observe that these must be consistent with our being here to do the observing. John Barrow and Frank Tipler give this formal expression:

> The observed values of all physical and cosmological quantities are not equally probable but they take on values restricted by the requirement that there exist sites where carbon-based life can evolve and by the requirement that the Universe be old enough for it to have already done so.[19]

The Weak Anthropic Principle amounts to little more than tautology. 'We're here and so things are the way that makes that possible.' It fails adequately to encapsulate the remarkable degree of 'fine-tuning' involved in spelling out the conditions that have permitted our evolution. Only a tiny fraction of conceivable universes could have been the homes of conscious beings.

Much too strong as a *scientific* principle is Barrow and Tipler's formulation of the Strong Anthropic Principle:

> The Universe must have those properties which allow life to develop within it at some stage of its history.[20]

Where could such a necessity originate from within science alone, if that discipline has foresworn any consideration of purpose? The Strong Anthropic Principle is frankly teleological in its insistence that the world 'must' have been that way.

A philosophical parable

In my restrained English way, I have suggested a Moderate Anthropic Principle:

> which notes the contingent fruitfulness of the universe as being a fact of interest calling for an explanation.[21]

We are back with Leslie's fly on the blank wall. Why did the shot hit its apparent target so accurately? We should not merely shrug our shoulders and say it just happened that way. Actually we need to enhance the story somewhat. Let us replace the not intrinsically important fly with a tiny button, which on its being struck opens up the door of a secret treasure house. It is not long odds that make an event particularly notable, but only their combination with some other source of meaning that then makes that particular event the carrier of significance. Any specific layout of small white stones on green grass is hugely improbable, because there are so many possible configurations in which they might lie. It is only when there is something extra associated with such a pattern – such as its forming the letters SOS – that we think an explanation is called for. The evolution of conscious life seems the most significant thing that has happened in cosmic history and we are right to be intrigued by the fact that so special a universe is required for its possibility.

It is interesting how resistant some scientists are to seeking wider understanding. Heinz Pagels criticized the Anthropic Principle because it is not subject to experimental falsification, 'a sure sign that it is not a scientific principle'[22]. One wonders if Pagels regarded the theory of evolution as as scientific principle and what was the experiment he thought would falsify it. When he goes on to say 'I would opt for rejecting the anthropic principle as needless clutter in the conceptual repertoire of science'[23], he displays a sad reluctance to lift his eyes beyond the horizon of the most narrow construal of scientific knowledge. We do not need to condemn ourselves to so impoverished a view. The question of the significance of the Anthropic Principle is a scientific *metaquestion*; that is to say it arises from the insights of scientific cosmology but it goes beyond what science alone is competent to discuss. We are concerned here, not with physics, but with metaphysics – and that is as true of those who deny a significance to the Principle as it is of those who wish to seek for a more profound understanding. In addressing questions of purpose or non-purpose, they are going beyond the self-limited domain of scientific discourse.

Leslie's philosophical story indicates the two possible lines along which a deeper intelligibility might be found. The fly was hit *either* because very many bullets were hitting the wall *or* because a marksman had taken careful aim.

The first explanation translates into a many-universes understanding of anthropic significance. If there are many, many different universes, each with its own physical laws and circumstances, then, somewhere in that vast portfolio of realized possibilities, there might well, 'by chance', be a world with just the right conditions for the evolution of carbon-based life. That is the universe in which we live, of course, because we could not have appeared in any other. If you fire enough bullets, one of them may just happen to hit the fly.

Many universes

The first thing to ask about this explanation is whether it is offered to us as physics or as metaphysics. One physical way of realizing a variety of different sets of effective laws of nature is provided by the concept of spontaneous symmetry breaking. When our universe was very hot and energetic, immediately after the Big Bang, it is supposed that the operative laws of nature were those of the (hypothesized) highly symmetrical Grand Unified Theory (GUT). As expansion cooled the universe, the symmetries present in the GUT were broken and the laws of nature as we now observe them crystallized out in the course of this process. Some of the details of the laws that we actually experience (including precise values of their force constants) depend upon the unpredictable way in which that symmetry breaking actually took place. It did not have to occur in a literally universal way. There could be different cosmic domains in which the details were different and, in consequence, the effective forces of nature would be different. It is supposed in this account that our experience of the universal character of natural law in that part of the universe where we are able to observe it is due to its domain wall (its interface with its differing neighbours) having been blown far away by the process of inflation.

If this speculative account is correct, it describes a universe that is a mosaic of (vast) sub-universes, in each of which a different set of force constants would be operating. These many sub-worlds would play the role of the bullets in the parable. We live in that domain where it has been possible for us to evolve. If true, this picture would go some way towards explaining anthropic coincidences. However, there would still be significant anthropic conditions to be satisfied by the universe as a whole. For example, its basic GUT would have to be such that under spontaneous symmetry breaking it could yield *appropriate* effective laws of nature. Not all GUTs would have this property. Thus, a total explanation of the Anthropic Principle would still call for an ensemble of different universes (in the large sense) and not just for an array of domains or sub-universes.

Other speculative attempts at a 'physical' explanation clearly go beyond science itself. This is true of the proposal that the universe has been undergoing a sequence of expansions followed by collapses, a chain of Big Bangs followed by Big Crunches, and that from each crunch it bounces back with totally different physical laws. The singularity of Big Crunch/Big Bang is inaccessible to scientific thought and it is pure metaphysics to postulate that it wipes the physical slate clean and enables a quantitatively different universe to emerge.

Others have appealed to the highly contentious 'many-worlds' interpretation of quantum theory.[24] This supposes that the different outcomes of a quantum measurement all actually occur, but in disjoint worlds into which physical reality splits at each such act of measurement. Such speculative prodigality has appealed to comparatively few physicists, but even if it were true, the result would be the generation of worlds with different histories but not universes with different laws of nature. The many worlds differ only in the consequences of measurement, not in their physical fabric.

Thus, even after exploiting the ambiguities of spontaneous symmetry breaking, a full many-universes explanation of anthropic coincidences is metaphysical in character, depending upon an appeal to the existence of worlds of whose being we can have no direct, scientifically motivated, knowledge. It is a metaphysical guess that they might be there.

Creation

It is also a metaphysical guess that there might be a God whose will and purpose is expressed in a single universe, endowed by its Creator with just the physical fabric that will permit it to enjoy a fruitful history. This, of course, is the appropriate translation of the marksman solution to the problem of the shot that hit the fly on the wall. There was a purpose behind that careful aim. Notice that this theistic understanding of anthropic fruitfulness contains a complete answer to the anthropomorphic criticisms of David Hume, directed against less sophisticated accounts of creation. God's conferring on the natural world the power to make itself through an evolving history realizing that endowment of fertility is as far as could be from the image of human craft engaged in moulding pre-existing material. This new natural theology is not presented as a rival to science, as if it were an alternative explanation of the process of the world, but as a complement to science, making more deeply intelligible those fine-tuned natural laws that science must *assume* as the general basis of its explanations of particular occurrences. (We can note that the other main criticism made by Hume – that the existence of suffering in the world shows 'creation' to be very imperfect – can at least be addressed by the insight that this is the necessary cost of a universe allowed to make itself, whose shuffling explorations of possibility will have to have ragged edges.[25])

How are we to judge the matter? Leslie ends his account of the issues with the conclusion.

> My argument has been that fine tuning is evidence, genuine evidence, of the following fact: *that God is real and/or there are many and varied universes.* And it would be tempting to call the fact an observed one. Observed indirectly, but observed none the less.[26]

In that 'and/or' he indicates his judgement that either metaphysical interpretation is equally likely. If that were all to be said, I would agree with him. But, of course, I believe there are many other arguments for belief in God – including those from the intelligibility of the physical world and from religious experience – and if that is the case then the anthropic considerations are but part of a cumulative case for theism.[27] Thus, I believe that in the delicate fine-tuning of physical law, which has made the evolution of conscious beings possible, we receive a valuable, if indirect, hint from science that there is a divine meaning and purpose behind cosmic history.

In my opinion, science is possible, and cosmic history has been fruitful, because the universe we inhabit is a creation. In fundamental terms, that is what it all means.

NOTES

1. Quoted in R. S. Westfall, *Never at Rest* (Cambridge University Press, 1980), p. 748.
2. Quoted in C. E. Raven, *Science and Religion* (Cambridge University Press, 1953), p. 210.
3. R. Dawkins, *The Blind Watchmaker* (Longman, 1986), p. 5.
4. Quoted in H. Montefiore, *The Probability of God* (SCM Press, 1985), p. 159.
5. Quoted in J. H. Brooke, *Science and Religion* (Cambridge University Press, 1991), p. 314.
6. P. Teilhard de Chardin, *The Phenomenon of Man* (Collins, 1959).

7. A. R. Peacocke, *Creation and the World of Science* (Oxford University Press, 1979).
8. See also J. Monod, *Chance and Necessity* (Collins, 1972).
9. J. C. Eccles, *Evolution of the Brain* (Routledge and Kegan Paul, 1984), ch. 2.
10. P. Davies, *The Cosmic Blueprint* (Heinemann, 1987), p. 151; cf. F. Hoyle, *The Intelligent Universe* (Michael Joseph, 1983).
11. Davies, *Blueprint*, p. 203.
12. J. C. Polkinghorne, *Science and Creation* (SPCK, 1988), ch. 4; *Science and Christian Belief* (SPCK, 1994), ch. 4.
13. J. C. Polkinghorne, *Reason and Reality* (SPCK, 1991), ch. 3; 'The Laws of Nature and the Laws of Physics' in R. J. Russell, W. R. Stoeger and G. V. Coyne (eds.), *Quantum Cosmology and the Laws of Nature* (Vatican Observatory, 1993), pp. 437–48.
14. J. D. Barrow and F. J. Tipler, *The Anthropic Cosmological Principle* (Oxford University Press, 1986); J. Leslie, *Universes* (Routledge, 1989).
15. Quoted in Barrow and Tipler, *Principle*, p. 22.
16. See note 14; also Polkinghorne, *Reason*, ch. 6.
17. Leslie, *Universes*, pp. 17–18.
18. P. Davies, *God and the New Physics* (Dent, 1983), p. 179.
19. Barrow and Tipler, *Principle*, p. 16.
20. *Ibid.*, p. 21.
21. Polkinghorne, *Reason*, p. 78.
22. H. Pagels, *Perfect Symmetry* (Michael Joseph, 1985), p. 359.
23. *Ibid.*
24. See J. C. Polkinghorne, *The Quantum World* (Longman, 1984), ch. 6.
25. See Polkinghorne, *Creation*, ch. 4; *Science and Providence* (SPCK, 1989), ch. 5.
26. Leslie, *Universes*, p. 198.
27. Cf. R. Swinburne, *The Existence of God* (Oxford University Press, 1979).

7

Ultimate questions

It would be disingenuous not to acknowledge that a big question mark hangs over the claim of cosmic meaning set out in Chapter 6. Its origin is this. The universe we inhabit today is the result of fifteen billion years of fruitful evolution. But how will it all end? The honest scientific answer is 'badly' – in the collapse or decay of the whole cosmos. Such a tale of apparent futility led the distinguished theoretical physicist, Steven Weinberg, to write,

> It is very hard to realize that this all [the beauty of the Earth] is just a tiny part of an overwhelmingly hostile universe. It is even harder to realize that this present universe has evolved from an unspeakably unfamiliar early condition, and faces a future extinction of endless cold or intolerable heat. The more the universe seems comprehensible, the more it also seems pointless.[1]

This is a serious challenge, which faces those of us who take a theistic view of reality. In meeting it, we must first consider the detail of science's prediction about the future.[2]

The end of the world

Take life on Earth first of all. Our sun has been shining steadily for five billion years and it will continue to do so for about five billion years more. Then it will have burnt up all its hydrogen fuel and it will enter into the next stage of stellar evolution. This will turn it into a grossly swollen red giant, extending beyond the Earth's orbit in the solar system and burning up any life remaining here to a cindery frazzle, before recollapsing into a white dwarf. What we may fear might be brought about in the shorter term by human failing or mismanagement will certainly be brought about in the longer term by stellar explosion. Of course, five billion more years is a long time for further developments to take place and by then our successors may well have moved out of the solar system to found colonies on planets circling younger and temporarily less dangerous stars.

But what about the universe itself? What does its future hold? From the cosmic point of view, history is a gigantic tug of war between two opposing principles. One is the expansive force of the Big Bang, continuing to propel matter apart. The other is the contractive force of gravity, seeking to pull matter together. They are very evenly balanced and our present knowledge does not permit us to predict which will win in the end. Accordingly we have to construct two alternative scenarios for the universe's future. If expansion prevails, the galaxies now flying apart will continue to do so for ever. Within each of the galaxies, gravity will score a local victory, causing them to condense into gigantic black holes. These will then decay, over immense periods of time, into low-grade radiation. That way the universe ends in a long drawn out, dying, whimper. It is not a very encouraging prospect. Unfortunately, if gravity prevails the outlook is equally bleak. One day, the present expansion of the galaxies will be halted and reversed. They will come flying back together and what began with the Big Bang will end in the cosmic melting pot of the Big Crunch. That way the universe ends in a hectic, fiery, bang. Either way, in the end all is condemned to futility. It is as sure as can be that humanity, and all forms of carbon-based life, will prove a transient episode in the history of the cosmos. It will not happen tomorrow, of course. Tens of billions of years will elapse before one or other of these dismal destinies is realized.

Life's destiny

In the end then, what is the fate of intelligent consciousness to be? Who will care for it? Amid the gloom, there seem to be two possible, more optimistic, answers. One is 'Life itself'; the other is 'God alone'.

Those who give the first answer argue like this. Life has come to be through the evolved complexity permitted by the elaborate and fruitful chemistry of carbon. However, when life reached the stage of human self-consciousness the evolutionary process was radically modified. Natural selection is not given free reign, for human compassion preserves and provides for the weak and the disadvantaged. Above all, culture affords a much more powerful and quickly effective way of transmitting information from one generation to the next than that provided by genetic channels based on DNA. At present, culture (which of course includes science and technology) is simply ancillary to the development of life, giving us means to achieve ends that we could not accomplish unaided. Eventually may not culture go beyond that and create new, artificial, forms of 'life' itself? Those who make strong claims for artificial intelligence, and who believe that truly 'thinking' computers are an inevitable future development, are prophesying that a form of silicon-based life will be humanly created, to take its place alongside the carbon-based life of its creators. Once such a process has begun, it will surely continue. As cosmic circumstances change, either in the freezing of a dying expanding universe or the frazzling of a fiery collapsing universe, so that any form of electrochemical life is no longer possible (whether carbon-based, or silicon-based, or whatever), will not life engineer further (and to us stranger) reimbodiments of itself, taking advantage of whatever purchase it can gain on the increasing hostile cosmic circumstance? In that way life will defy the threat of extinction and preserve itself as long as the universe lasts.

The strongest form of these claims is asserted by what John Barrow and Frank Tipler call the Final Anthropic Principle: 'Intelligent information-processing must come into existence in the Universe, and, once it comes into existence, it will never die out'.[3] Tipler, in particular, has pursued the matter with a peculiar tenacity.[4] He treats the essence of life as being the processing of information. This would imply that there would be an infinite degree of fulfilment in the course of cosmic history if there could be the processing of an infinite number of bits of information. Tipler's conclusion is that, in certain particular and specified circumstances, this would be possible in the hectic period of the final split seconds of a collapsing cosmos. The whole universe would then have been taken over by 'life', in an ever-more-energetic, ever-faster racing, computer-like mode, which in its dying gasp would achieve an infinity of information processing operations. The energy for this operation would be derived from the gravitional shear energy of cosmic collapse. Tipler explicitly compares this ultimate cosmic computerization to the attainment of Teilhard de Chardin's Omega Point[5]. Tipler is a kind of Southern Baptist atheist and he speaks of his ideas as 'physical eschatology', and of the fleeting ultimate state of his universe as the realization of a 'physical god'.

It is a strange, not to say bizarre, vision of the cosmic future. There are considerable difficulties concerning its claims. The first is that it is based on a computer model of the nature of life. Living entities are seen as being finite-state machines and the character of life is the processing of information. Tipler's physical god is the apotheosis of artificial intelligence. I cannot accept so physicalist, reductionist, a view of life. The claims of artificial intelligence appear to me to be inflated and implausible. Thought is more than the execution of algorithms; computers can do the latter but not the former (see Chapter 5).

Another difficulty is to be sure that, even if it were feasible, life would in fact choose to engineer its own continuance. The ambiguities of human history would suggest that self-destruction might be an equally plausible possibility. Tipler seeks some guarantee for the coming-to-be of Omega by imposing a future boundary condition on his version of the quantum cosmology describing the universe. This ensures that all parts of the cosmos are in causal contact with each other in those closing instants of collapse. This is a necessary condition for the possibility of Omega, but not a sufficient condition. My doubts are not resolved by the argument.

A third difficulty relates to the wealth of conjectures Tipler has to make about how matter will behave in the ultra-energetic final regime of a collapsing universe. Yet another blank cheque is being drawn on a totally unknown physical account. The speculations of writers such as Stephen Hawking about the very early beginnings of the universe pale into insignificance when compared with Tipler's confident assertions of what will happen at its ending, involving behaviour in regimes utterly remote from the reach of motivated conjecture.

Thinking about life's self-perpetuation in the course of a proposed physical eschatology reminds one of the ancient tale of human presumption in building the Tower of Babel. In both cases an ambitious ability is being claimed for creatures that might more properly be thought to belong to their Creator. It is time to explore the alternative answer and to consider the possibility that it is to God alone that the ultimate care of life belongs. The attempt to inquire into whether such a view makes sense inevitably involves me in a greater appeal to theological understanding than I have allowed myself to make elsewhere in this book.

The true ultimate

For the religious believer the only true ultimate is God himself. He is the 'bottom line' in any discussion of the significance of past, present or future. If there is an ultimate hope, an intimate and unfailing care for creation, it can be found only in the eternal faithfulness of the Creator.

That was the point Jesus made in an argument about whether there is a human destiny beyond death. The Sadducees did not believe that there was. They based their religious belief on the *Torah*, the first five books of our Bible, and they did not think they could find such a hope set forth in its pages. They came to Jesus with a rather ingenious conundrum about a woman who had successively married seven brothers, asking whose wife she would be in the age to come? As he so often did, Jesus cut through the superficial argument and went straight to the real point. He recalled an incident in the *Torah*, where God speaks to Moses from the burning bush and reminds him that the Lord is the God of Abraham, the God of Isaac and the God of Jacob. Jesus goes on to say 'He is not God of the dead, but of the living' (Mark 12:27). In other words, if the patriarchs mattered to God once, as they certainly did, they matter to him for ever. He will not cast them away like broken pots once they have served his immediate purpose. The Lord is eternally faithful in his care for Abraham, Isaac and Jacob, for you and me. And, I am sure we can add, for all of creation in appropriate ways. I will return to this last point shortly, but first let me consider whether in this scientific age we can seriously believe that there is an ultimate destiny for human beings beyond their death.

The first question we have to ask is what is the nature of men and women? I have already explained (Chapter 5) that I think we are psychosomatic unities and that the soul is the 'form' (the information-bearing pattern) of the body. That pattern will be dissolved at death with the decay of my body. Yet it seems to me to be a perfectly coherent hope that the pattern that is me will be remembered by God and recreated by him in some new environment of his choosing in his ultimate act of resurrection. The Christian hope has never truly been that of the survival of a spiritual component. We are not apprentice angels, longing to rid ourselves of the encumbrance of our bodies. Rather, it is the essence of humanity to be embodied. The Christian hope, therefore, is of death and resurrection, an end and a new beginning.

The 'software' of life cannot run on any old 'hardware' (there is something essential in our specific embodiment, I believe, and I reject a strict computer analogy) but there is surely the possibility of there being a new 'matter' in which we can be re-embodied at our resurrection. Where will this 'matter' come from? I think that it will be the transformed matter of this present world, delivered by God from its destiny to decay. The universe is going to die but, because God cares for it, it will have its resurrection beyond its death, just as we shall have our resurrection beyond our deaths. In fact the two destinies are one. This is the Christian hope, articulated by St Paul, that there is to be a new creation, the redemption of the old creation from its frustration and futility (Romans 8:18–25; 2 Corinthians 5:17), in which we shall all share.

It is a wonderful, deeply mysterious and exciting, vision of hope. Two questions press upon us. What reason do we have to think it might be true? Does it make sense to believe in a new creation?

I have already given the fundamental reason for a hope that cosmic collapse or decay is not the last word. It is a trust in the everlasting faithfulness of God the Creator, who will not allow anything of good to be lost. For humankind that implies an individual destiny for each of us. We are all going to die with our lives incomplete, our hurts unhealed, business unfinished, potentialities unexplored or unfulfilled. I do not think that this life by itself makes sufficient sense to be considered complete without the possibility of a further life and growth beyond. The fundamental question around which the discussion of this chapter revolves is the question of whether the universe makes total and absolute sense – not simply the wonderful rational sense that science has discovered the physical world to possess, but an ultimate sense that embraces the whole range of human insight and aspiration – or whether it is a tale 'told by an idiot'? In the end, the more we comprehend the universe, does it become more pointless or become more truly a cosmos, totally meaningful to us so that we are truly at home in it and not lone protestors against its absurdity? My instinct as a scientist is to seek a comprehensive understanding and I believe that it is my religious faith that enables me to find it.

There is a final point that candour requires me to add. The Christian has another reason for believing in an ultimate destiny beyond the immediate fate of physical death. It lies, of course, in the belief in the resurrection of Jesus Christ. This is not the place to explore further the historical motivation for this hopeful belief, or to defend its reasonableness within a theistic understanding of reality. That is a task I have attempted elsewhere.[6] The truth of Christ's resurrection is, for me, a truly ultimate question.

NOTES

1.   S. Weinberg, *The First Three Minutes* (André Deutsch, 1977), pp. 148–9.
2.   For a fuller account of what follows, see J. C. Polkinghorne, *Science and Christian Belief* (SPCK, 1994), ch. 9.
3.   J. D. Barrow and F. J. Tipler, *The Anthropic Cosmological Principle* (Oxford University Press, 1986), p. 23 and ch. 10.
4.   F. J. Tipler, 'The Omega Point Theory: A Model of an Evolving God', in R. J. Russell, W. R. Stoeger and G. V. Coyne (eds.), *Physics, Philosophy and Theology* (Vatican Observatory, 1988); F. J. Tipler, *The Physics of Immortality* (Macmillan, 1995).
5.   P. Teilhard de Chardin, *The Phenomenon of Man* (Collins, 1959), Book 4, ch. 3.
6.   See J. C. Polkinghorne, *Belief*, chs. 6 and 9.

8

Is, ought and wonder

Value free?

It is quite customary to speak of science as being 'value free'. Several things might be meant by that statement. One would be that scientific argument does not make an appeal to value in setting out its case. It does not say 'This is the way things ought to be and so they must be that way'; it simply says 'This seems to be the case'. 'Is' and 'ought', we are told, are quite distinct categories and science restricts itself to the former.

What I have said certainly fits the way scientists write their papers for publication in the learned journals, but it does not correspond to the way scientists talk among themselves. 'It must be right' is the way they feel about an elegant and insightful idea, often long before the empirical adequacy of the theory has been verified to a degree sufficient to warrant such a conclusion. Of course, such feelings are not invariably correct, but they are confirmed astonishingly frequently in our experience. We have already noted (p. 15) that Einstein believed that his idea of special relativity was just too good to be wrong. When his experimental critic, D. C. Miller later pressed his case, Einstein wrote to a friend 'I have not for a moment taken [his results] seriously'.[1] Subsequent developments amply confirmed Einstein's confidence (though no one has succeeded in explaining what was the source of Miller's mistake). This instructive story does not quite correspond to the austere official notion of 'value-free' argument.

A second meaning of 'value-free' in relation to science would be that the description of the world that it offers us is one in which there is no lodging for the idea of values. Scientific discourse is concerned with a physical world whose processes involve exchanges of energy. Its language is the quantitative account of matter and motion, not the qualitative account of beauty and morality. Things just happen and the question of value and meaning in what is going on is deliberatively bracketed out. This is the great methodological programme of Galileo and John Locke: to concentrate on the primary characteristics of measurable quantities and to neglect the secondary characteristics of humanly perceived qualities. As an investigative strategy, focusing attention on limited but attainable goals, it was a brilliant move. It would be a great mistake, however, to confuse this heuristic device with an adequate approach to reality as a whole. Methodology does not determine ontology. A projectile executes the same parabola under gravity, whether it is a shell, or a distress flare, or the signal to attack, but that does not mean that Newtonian physics by itself is an adequate account of what is going on. The fact that meaning and value have been bracketed out by science does not imply that meaning and value do not exist. Those who deliberately shut their eyes cannot thereby deny the reality of vision, any more than seventeenth century savants denied the existence of mountains on the moon by declining to look through Galileo's telescope. Moreover, the discovery of quantum theory shows that the subatomic world, and its associated primary quantities, cannot be surveyed with the degree of objectivity and complete detachment that classical physics had assumed. It is a matter of unresolved dispute exactly what is the extent of interaction between observer and the system observed that is involved in the act of measurement[2], but it is widely supposed that quantum theory implies a degree of observation-influenced reality, even in the limited domain of science.

While the description of value is absent from the account of science, it is certainly not absent from the scientific method itself. In Chapter 2, we noted the underdetermination of theory by experiment, in the sense that theories claim to describe an infinity of different circumstances, whilst experiments can only probe a finite number of particular cases. In practice, however, this does not cause paralysis of thought, as scientists haver in the face of a bewildering multiplicity of options. Far from it! In fundamental physics the difficulty is precisely the reverse. Long and arduous is the search for a single acceptable theory. The satisfaction of criteria involving value judgements about simplicity and naturalness is essential for a theory's acceptance. Contrivance and complexity are fatal flaws. Physicists have learnt that truly successful theories are always characterized by their being expressible in terms of 'beautiful' mathematical equations (pp. 78–80). Three hundred years of the experience of doing fundamental physics supports this conclusion, so that this search for beautiful equations is more than a mere mathematical aestheticism. The reason that we believe that we find the best explanation of physical phenomena in this way derives from our experience that such theories have time and again proved to have a fruitfulness extending far beyond the original phenomena for which they were invented. In science, the beautiful is the good because it has proved to be the fertile. Dirac's lifetime search for beautiful equations is an object lesson that this is so, as is Einstein's discovery of general relativity through a similar eight-year quest.

Such uncovenanted fruitfulness impels the conviction that scientific theory is on to something, that these beautiful equations do indeed describe a true aspect of reality. Their existence corresponds to another value-laden aspect of the scientific life, the experience of wonder at the deeply satisfying structures of the physical world revealed to our inquiry. Here is the reward for all the weariness and frustration that are inescapable components of any serious scientific investigation, as in any other kind of worthwhile activity. In our human nature, not only has the universe become aware of itself, it rejoices in that awareness.

A moral community

The whole practice of science depends upon the acceptance of certain moral values within the scientific community: honesty in the reporting of results, respect for the credit due to others for their discoveries, generosity in making results available to one's colleagues. I do not wish to claim that scientists are all morally whiter than white, but cases of fraud or plagiarism are very rare, and correspondingly shocking when discovered. Theoretical physicists do not normally clean their blackboards when they go home in the evening, but they leave their current thoughts written up in the equations they are studying, confident that someone will not sneak in and steal their ideas.

Beauty

We need to view the richness of reality with multiocular vision. Once we open more than our scientific eye alone, we behold a world that is flooded with value. A scientist could tell you all about vibrations in the air and analyse their frequences and energies, and give an account of the neural responses of the ear. Having said that, as a scientist that person would have nothing more to say. But as a human being, he or she would recognize that no real contact had yet been made with the mysterious reality of music. That experience requires a different form of perceptual encounter. It seems crass to dismiss music as no more than an epiphenomenal ripple on an essentially silent reality. Many scientists feel a particular affinity with various kinds of music, particularly contrapuntal forms whose interweaving patterns are reminiscent of the subtle dynamical structures of the natural world. Is music's profound ability to speak to us of eternity through a temporal succession of sound, to be dismissed as a discardable 'secondary quality'? Is the true reality of a Rembrandt self-portrait just the chemical composition of its specks of paint? It seems absurd to say so. What an implausibly diminished view of reality one would embrace if one were to confuse the scientific methodology of Galileo and Locke with a statement about the ontological nature of the universe. Yet, there is a deep mystery in our auditory and visual encounters with beauty. Our experience is constrained by our neural responses to stimuli. We can only see light of a certain range of wavelengths, only hear sounds of a certain range of frequency. Our window of perception is limited, but it appears to be wide enough to let reality into our minds. I cannot dismiss our experience of beauty as just an epiphenomenal spin-off from the 'hard wiring' of our brains. Something of true significance is being encountered. As with our scientific ability to comprehend both the spacious universe and the quantum microworld, we appear to possess remarkable powers of access to reality. It is in the exercise of such powers that the deepest satisfactions of human life are found to lie.

A comprehensive view

An arid world, devoid of value, would, as we have seen, be one in which the practice and experiences of science itself would find no home. Its impersonal setting could not contain the personal subject who is the necessary knower of scientific knowledge. The theologian Langdon Gilkey comments,

> In scientific realism as a worldview, the knowing subject is reduced to a known object, consciousness to neurology, the ordered cosmos of science to a system of inert parts, all without the organizing mind of the scientists, the genius of the neurologists, the creativity of the knowing subjects – who, incidentally, are performing the reduction.[3]

He says later, 'Not all that we know is science, lest there be no possibility of science'.[4] Scientific discovery is the activity of persons. It can be assisted by computers but it could not be delegated to them. One cannot divorce personhood from the experience of value, those tacit judgements of fittingness and economic elegance that are a central part of the creative discovery of scientific theory.[5]

I do not for a moment believe that the marvellous order science discerns is an order that scientists impose upon the flux of experimental experience. The Kantian notion that things in themselves are totally inaccessible to us and that all we can do is filter phenomena through the grid of a prior expectation, whose pattern is then stamped upon them, seems to me to be entirely contrary to actual scientific experience. It would imply that physical events are somehow so plastic in their character as to be capable of being twisted into whatever shapes please our intellectual fancy. Physics' encounter with the stuff of the universe has precisely the opposite character. The world proves recalcitrant to our expectation, stubbornly resisting what we expect of it and thereby forcing us painfully into new theories that surprise us and exceed in interest anything we could have imagined beforehand. Just think of quantum theory, or of the theory of *confined* quarks, sitting there inside nuclear matter but incapable of being ejected from it, however energetically bombarded. No one expected the physical world to be like that. Hence the feeling of *discovery*, so central to scientific practice. Of course, we never grasp the whole of physical reality – we have acknowledged already that verisimilitude (see Chapter 2), rather than absolute truth, is the best that science can claim – but we do achieve partial but reliable insight into the way things are. In an analogous way, I believe our experiences of beauty are not just human projections onto a neutral world, but they are engagements with a true component of reality.

Ethical values

The same holds for another form of human encounter with value, our ethical intuitions. Here it seems we have access to a form of knowledge different in its origin and character from scientific knowing. I believe that I know as certainly as I know anything that love is better than hate, truth is better than falsehood, that torturing children is wrong.

There are those who would respond to that claim by pointing to the culturally influenced element that is present in moral attitudes. 'You may think love is better than hate' they say, 'but what about the Ik tribe in Uganda, whose way of life appears to be based exclusively on selfishness and an implacable hostility to the outsider?'. The same cultural relativists might appeal to the Azande tribe as a refutation of scientific claims that the death of a chicken from 'benge' is to be explained on the biochemical basis of the action of a toxin. The tribe would attribute it to the oracular discernment of witchcraft. Such an interpretation fits coherently into the Azande world view. Who are we to correct the Ik or the Azande?

Well, I think we can take on both of them. Modern biochemistry can explain much more than the Azande can and we have every reason to prefer its mode of understanding when it comes to the action of benge. We may ask what our Western cultural relativists do when they are ill. Is it a matter of indifference to them whether they consult a witchdoctor or a medical doctor? I think not.

We can recognize the presence of cultural influences on our judgements of value without feeling that we are so in thrall to those influences that there is no escape from them and no grounds for discrimination about the judgements being made. I believe I have a defensible basis for supposing the Azande to be wrong about benge. I believe I have a defensible basis for the moral belief that the Ik are wrong in rejecting altruism. I am not condemned to believe that they see it their way and I see it my way and there is nothing more to be said. The Ik are morally mistaken and the very peculiarity of their position (which makes them widely known in the Western world) derives from its perceived deviation from received ethical norms.

I do not think that my belief that torturing children is wrong is just a socially agreed convention of my society. I believe it is an insight into the way things are. In saying that I recognize that there have been societies that have practised torture, even, I am sad to say, in the name of religion through agencies such as the Inquisition. My response to that would be to say that culture had distorted the moral judgements of such societies and that our repudiation of torture shows us, in that respect, to be their moral superiors. It is not a matter of indifference, or mere convention, which attitude we take to the use of torture. Even where grave moral mistakes were committed, it is often the case that their per-petration was intended for what seemed to those concerned to be a desirable end. The widespread recourse to religious persecution by all sides in Reformation times was inspired by the belief that, whatever suffering might be endured in this life, it was preferable to the eternal torments of hell, destined for those who held the wrong religious beliefs. I believe that such an assessment was based on a terrible misjudgement about the nature of God and of divine wrath. The questioning in the nineteenth century of the idea that a loving God could exact infinite punishment for finite sins, however dreadful, was an important theological advance. The worst is indeed the corruption of the best; religion is always in danger of distortion by the demonic. Yet it remains the case that we can rightly suppose that the rejection of torture (or slavery, to take another example) represents a genuine moral advance and not just a change of social fashion.

Sociobiology

If a social-construct theory of value does not succeed, perhaps it can be replaced by the even more drastic ploy of a reduction to biology. There are those who claim, contrary to what I have been suggesting, that human perceptions of moral and aesthetic values can be given an explanation in scientific terms. What we might think of as intimations of reality are, in fact, no more than responses programmed into us by the selfish genes.[6] These intuitions are simply disguised strategies for survival. Their origin lies, not in the way things are, but in the way history has been. They are the deposit in human nature of the consequences of the evolutionary struggle. So goes the thesis of sociobiology in its strong form.

People who feel that one grand principle is the explanation of all that is are always difficult to engage with in discussion, because everything is made into grist for their ideological mill. It is a familiar difficulty in dealing with Freudians, who can readily produce explanations on their own terms for any lack of belief on yours. A rather similar problem afflicts encounter with those who assign an almost omnicompetent role to evolutionary argument. If evolution is the explanation of what is, then, of course, anything that is the case must have an evolutionary explanation – that is to say, it *must* have some survival value. The self-fulfilling circularity of the argument will satisfy only those already convinced of its truthfulness. On investigation by the uncommitted, however, there seems to be considerable difficulty in substantiating strong sociobiological claims.

Firstly, there is the question of whether we have any reason to believe at all that there are genes which generate in the phenotype specific modes of behaviour or modes of perception: a gene for altruism or for musical appreciation, say. Given the subtlety of human motivation and perception, it seems most unlikely that this is the case. Blue eyes are one thing; the expression of artistic genius is another. The bizarre and distasteful dissections of the brains of remarkable people such as Lenin and Einstein have not led to any useful insight whatsoever. Of course, what we are is constrained in various ways by our physical constitution, which itself is constrained in various ways by our genetic inheritance, but there seems to be left a long leash permitting a great variety of possibilities for individual cultural and moral development. We scarcely look like genetic robots.

Secondly, the matters to be explained just do not seem open to evaluation in survival value terms alone. Only the most ideologically committed can really think that the human ability to understand the strange counterintuitive sub-atomic world of quantum theory, remote from everyday experience, is a by-product of our ancestors having had to struggle for existence. If survival value is the clue, why do we find so much beauty in the inhospitable landscape of the desert? A notorious problem for sociobiologists is the origin of altruism, the value assigned to compassionate self-sacrifice on behalf of the other, which finds a place in all the great moral systems of the world. J. B. S. Haldane was genetically correct when he said that he would lay down his life for two brothers or eight cousins. The calculations of gene survival endorse such a strategy. But we may suppose that he would, in fact, have been willing to risk his life to save a totally unrelated stranger from drowning. Faced with people trapped inside a burning house, would he first have made inquiry of their kinship before attempting a risky rescue? I think not. Moral imperatives are much more than genetic survival strategies.

Richard Dawkins has sought to augment the genetic reductive argument by a second line of discourse. He proposes the existence of ideological units, which he calls memes, which propagate in a competitive way in human society, and human psyches.[7] Here is a second, mental, string to the neo-Darwinian bow. While this is an amusing metaphor, it does not begin, in my opinion, to amount to an adequate description of the history of human culture. Only a vastly oversimplified account could see it as the warfare of discrete, identifiable, units of idea. A statement like 'The meme-complexes of Socrates, Leonardo, Copernicus and Marconi are still going strong'[8] is a pretty ludicrous way of talking about culture.

The problem with all reductive theories is their self-destructive character. If we are just gene-machines and meme-machines, human rationality itself is threatened. The sociobiologists, if they are to maintain their position, must grant *themselves* a tacit saving clause. Otherwise their conclusions are as much explained away in a merely reductive manner as would be the other aspects of human knowledge that they seek to subvert. Ultimately their strategy, if pursued with consistency, proves to be suicidal. The insights of science would prove as genetically determined as the insights of morality.

The universe as creation

It is time to draw this discussion to a close. No adequate account of reality could fail to recognize that we live in a world suffused with value. The rational beauty that science discovers in the structure of the universe, and the sense of wonder enjoyed by scientists in the face of such discoveries, is part of that encounter with value. A world made up of quarks and gluons and electrons is also a world that is the carrier of a deep aesthetic beauty and the arena in which we exercise moral choice in accordance with those ethical intuitions that are part of our knowledge of reality. As part of its own peculiar style of inquiry, science may choose officially to neglect this dimension of value, but such a methodological limitation does not warrant the conclusion of a corresponding ontological poverty. There remains the mystery of the multi-layered reality that we encounter, a world simultaneously orderly, beautiful, and moral. What ties it all together? A possible answer that is coherent and intellectually satisfying – I claim no more than that – is presented to us by theistic belief. Reality is multi-layered because it is a creation. Behind the scientific order of the universe is the Mind of the Creator. Behind the human experience of beauty is the Creator's joy in creation. Behind our intuitions of morality is the discernment of the good and perfect will of the Creator. I actually believe that the grandest Unified Theory, the true Theory of Everything, is provided by belief in God.

NOTES

1. Quoted by A Pais, *'Subtle is the Lord . . .'* (Oxford University Press, 1982), pp. 163–4.
2. See, for example, J. C. Polkinghorne, *The Quantum World* (Longman, 1984), ch. 6, and for a careful philosophical account, B. d'Espagnat, *Reality and the Physicist* (Cambridge University Press, 1989).
3. L. Gilkey, *Nature, Reality and the Sacred* (Augsburg Fortress, 1993), p. 16.
4. *Ibid.*, p. 39.
5. M. Polanyi, *Personal Knowledge* (Routledge and Kegan Paul, 1958).
6. R. Dawkins, *The Selfish Gene* (Oxford University Press, 1976).
7. *Ibid.*, ch. 11.
8. *Ibid.*, p. 214.

9

Responsible behaviour

What we consider to be a responsible encounter with the natural world will be determined by how we conceive of nature. If nature is enchanted – every tree contains a dryad, every spring a naiad – then we shall encounter its magicality with a wary and propitiatory caution. If nature is just the backdrop for the real drama of humanity, then we shall encounter it with indifference or manipulate it to fulfil our immediate purposes. If nature is the womb that gave us evolutionary birth, we shall encounter it with a grateful kinship. If nature is (as I believe) a creation, then we shall respect it as God-given and seek to care for it. I think it right in this chapter about ethically responsible behaviour to lay bare some of the Christian roots of my concern and understanding.

Exploitation

Despite this last assertion, many people have argued that Christianity has encouraged a disastrously exploitive treatment of the natural world. Professor Lynn White wrote that 'Christianity insisted that it is God's will that man exploit nature for his proper ends. Christianity has a huge burden of guilt'.[1] Our recognition of the ruthless pollution of the environment in Eastern Europe, and in the former Soviet Union, by regimes inspired by atheistic communism, shows that the blame to be assigned does not all fall upon Christianity alone. Nevertheless one must recognize that there has been a current in Christian thought that has taken a purely instrumental view of nature. John Calvin wrote that 'the end for which all things were created [was] that none of the conveniences and necessaries of life might be wanting to men'. Christian writers often took a low view of the nature of animals. Aquinas wrote that 'If any passage in Holy Scripture seems to forbid us to be cruel to brute animals, that is either lest through being cruel to animals one becomes cruel to human beings or because injury to an animal leads to the temporal hurt of man'. Such a chilling view could only be reinforced later by René Descartes' understanding of animals as mere automata.

Yet that is only part of the story. Christian thought has consistently stood out against any interpretation of the material world as inherently evil (against manichaeism) or of human beings as purely spiritual entities seeking release from the gross entrapment of the flesh (against gnosticism). A religion of the incarnation – 'the Word was made flesh and dwelt among us' (John 1:14) – has to accord the highest value to the created order, in which God's Son has been a participant. At the heart of Christian spirituality is the sacramental life, in which the material objects of water and bread and wine are understood as carriers of the divine grace and presence. It is a commonplace to describe Christianity as the most materialistic of the world's great religious faiths.

Caring for creation

It is not surprising, therefore, that there are alternative Christian traditions that express the value of the natural world and a comradely respect for it. One finds this articulated in St Francis' 'Canticle of the Sun', where God is praised for all his creatures, including 'brother sun', 'sister moon', 'brother wind', and 'our mother earth which doth sustain us'. The Celtic tradition mingles the biblical and the natural in its hymns of praise:

> Abraham, founder of the faith, praised you:
>
> . . .
>
> Let the life everlasting praise you,
>
> . . .
>
> Let the shorn stems and the shoots praise you.

This understanding leads to an encounter with nature in which humans are actors in the wide drama of creation, actors who have emerged from the scenery of nature and who retain an intimate connection with it. This is beautifully expressed in the ancient story of Genesis 2, in which God forms Adam from the dust of the ground. The natural world is not there as an insignificant backdrop to human activity but it is our home, to be cherished and treated with respect as the sustainable resource for living. Humankind's relation with nature is expressible in terms of a stewardly care (the tilling of a garden), in which men and women act as God's representatives and viceroys in their role of nature become aware of itself and aware of its Creator.

Some Christian writers, such as Philip Hefner, have wanted to go beyond this account and to speak of humanity as being 'created co-creators' with God.[2] It is certainly the case that the emergence of *Homo sapiens* has profoundly modified the form of the evolutionary *creatio continua*. The weak are no longer left to die, so that the operation of natural selection is significantly modified for human beings. Culture provides a much more powerful way of transmitting information from one generation to another than that provided by the natural selection of variations in genetic inheritance. This Lamarckian ability of culture to pass on learned experience has produced profound changes in the short period of the tens of thousand of years or so in which it has been actively operating. The new and ambiguous powers conferred by the discoveries of genetic engineering, permitting the interchange of genetic material between orders of being that could never be linked by selective interbreeding, represent the latest development in this process. Its potentialities for both good and ill will require the most careful and scrupulous scientific evaluation and ethical assessment. Certainly humankind is having a powerful impact upon nature. Yet I prefer the more modest language of stewardship to that of co-creators. Though certainly not intended by its proponents, the latter language can carry with it the danger of hubris. Humanity's root temptation is that, so powerfully expressed in the myth of Genesis 3, to attempt to 'be like gods'. To speak of ourselves as stewards acknowledges that duty of care which we owe to creation and it recognizes the Creator as the source of our power and opportunity.

The need for care and constraint in our encounter with nature is widely recognized. Even an apostle of free individual enterprise such as Lady Thatcher could give the salutary reminder that 'we do not hold a freehold on our world, but only a full repairing lease'. There is a long history of the recognition of the need for restraint in the human encounter with nature, of which, I suppose, the game laws' recognition of the necessity for a close season to permit breeding and the replenishment of stock would be an early example. The sporting tradition that one does not shoot a sitting bird is also a recognition that there are proper limits, arising from an innate respect for the creature, on the ruthlessness with which humans can interact with their environment and its inhabitants.

Animal rights

Nature is a complex reality and our understanding of our relationship with it will have to be correspondingly complex, diverse and nuanced. We have natural enemies. A good deal of shrewd common sense, critical of too undifferentiated and sentimental an encounter with nature, is enshrined in the couplet:

He loveth all, who loveth best –
The streptococcus in the test!

We surely do not think it was wrong to seek to eliminate the smallpox virus through a world-wide programme of successful inoculation. Our attitudes to nature are sometimes unduly influenced by the pathetic fallacy (the attribution to animals of all human powers of feeling, in particularly an anticipatory anguish about the possibility of danger, in addition to an immediate reaction to its presence) or by cultural arbitrarinesses (you can do more or less what you like to rats, but the bushy tails of squirrels make them altogether different). Some, like Andrew Linzey,[3] have wished to assign to animals rights broadly equivalent to those assigned to human beings. This is defended as being a God-given aspect of creation and opposition to it is sometimes derided as being speciesism. Why should *Homo* think it is different from any other animal genus?

There seem to me to be grave difficulties with this extreme point of view. Firstly, there is the question of demarcation. Few would assert similar rights for the worm or the ant; even fewer for the human immunodeficiency virus (HIV). Those who reject species-ism seem to embrace a kind of mammalism without too much difficulty. Even if one relies on sentience, it is still unclear where to draw the line. Then there is the point that human beings do seem to possess emergent properties of a kind without parallel elsewhere in the animal world. In addition to self-consciousness, there is the fact that we are moral beings. Those who speak in strong terms of animal rights do not also speak in similar terms of animal duties. That is because animals are not morally responsible creatures in the way that we are. There is a human moral duty towards the animal world, but I think that is best expressed in terms of a respect for fellow creatures and not in terms of the rights of fellow moral beings.

Sustainable life style

The World Council of Churches has in recent years been sponsoring an ongoing discussion, and to some extent a continuing campaign of action, under the title 'Justice, Peace and the Integrity of Creation'. Within itself, that title encapsulates some of the dilemmas and perplexities we face in evaluating the human encounter with nature. Justice requires that all the peoples of the world should have access to a fair share of its resources, and ultimately peace can only be founded upon such an equitable basis. That consideration applies not only to those alive today but also to those who are to come after us, so that there is the implication of a sustainable form of the use of the Earth's resources. Yet, for poor peoples living in the environment of the rain forests, such access may in reality only be immediately available through the felling of trees for firewood or in the effort to produce land for the cultivation of crops. In the longer term, these actions will continue the alarming rate of loss of precious natural resources, essential for global ecological stability and the preservation of biodiversity. In the longer term also, the needs of these indigenous peoples ought to be met through a fairer economic system embracing North and South, the rich and the poor countries of the world, but no one

should underestimate the immense international political difficulties in turning that aspiration into a reality. It would surely
require considerable willing sacrifice by the world's rich people –
that is to say, by *all* the inhabitants of Europe and North America.
It is easy to speak of living simply so that others may simply
live; it is immensely hard to implement that in some real and effective way that goes beyond a personal gesture. That is why people
of moral awareness and ecological concern should be found quite
as much in organizations such as the World Bank as in organizations such as Greenpeace.

The integrity of creation

It is also perplexing to understand truly what is meant by the integrity of creation. Too often it seems to be just a slogan for conservation pure and simple, an endorsement of a policy of no change.
How could that be possible in an evolutionary world? People often
show no recognition of the strangeness, indeed the ambiguity, of
the lessons which we might draw from looking at nature. We
should not look at nature only through human spectacles but recognize it in its otherness.

We are worried about genetic loss when the destruction of the
rain forest brings about the extinction of many of the species found
in its rich flora and fauna. I do not suggest that we should be
complacent about this, but it is a fact of evolutionary history that
times of taxonomic depletion and environmental change have also
been times for the opportunity for new developments in the history
of life. The decease of the dinosaurs gave the chance for the mam-
mals to take over. Of course, the timescales for reaping the fruits
of such changes are long compared with human historic memory
and presently we are exterminating species at ten thousand times
the rate of natural loss. However, we cannot simply conclude that
all we must do is to struggle to maintain the status quo.

At a more humble and particular level, respect for the integrity
of creation must surely imply that human actions should not frus-
trate the natural character and way of life of animals, for which

we have a duty of care. Modern factory farming methods often seem to go against this moral necessity in quite unacceptable ways. The keeping of chickens, pigs and calves in confined circumstances, so that they have no freedom of movement or means of rest, is surely wrong. Cheap food can be purchased at too great a cost in terms of animal welfare.

I do not think our relations with animals are all of one kind. Respect for integrity seems to me to imply that there are some animals with which our relationship will, in fact, have an adversarial character. There seems to be a natural form of hunting, whether for food or fur or for the elimination of predatory pests, in which those who undertake it show a real respect for their quarry. Of course, there is a need to avoid the inflicting of unnecessary suffering, but I believe that this is widely recognized by those people for whom hunting is a way of life and not just a townee's rapacious expedition into an alien countryside. I am influenced in this thought by my maternal grandfather. He was a head groom and often rode to hounds as a second horseman and sometimes as a whipper-in. He was a gentle fulfilled man, with a genuine and deep understanding of animals, both of the horses he rode and of the foxes he hunted. I think we have to be careful not to be too sentimental in our encounter with nature.

I also believe that it is acceptable, in a carefully controlled way, to use animals experimentally for scientific and medical research. Of course there must be stringent regulations to avoid unnecessary suffering and to ensure that justifiable and worthwhile purposes are being pursued. It is one thing to use animals to test a new antibiotic; quite another thing to use them for the routine checking of cosmetics.

In much of our encounter with the animal world it is the type, rather than the individual, which seems to be the focus of our respect. By that I mean that we wish to preserve deer, but if that requires the culling of a herd, most of us find that acceptable provided it is carried out in a humane fashion. Of course, that judgement is modified in the case of animals (such as pets) with which we have established an individual relationship. But I think that enhanced status then derives from the human side of the bonding, and we still feel able to have the animal put down if that ends otherwise unavoidable suffering.

The environment

It is time to lift our eyes to wider vistas and consider the human encounter with the environment as whole. Since culture began, there has been a continual impact of human beings upon their natural surroundings. In Britain, all the scenery has been shaped by the hand of human beings. Even what we regard as wild countryside – such as the fells of the English Lake District – has been created by the timber-cutting of our ancient ancestors. In other parts of the world there is still some true wilderness 'untouched by human hand', but Britain has none. That does not mean that we do not have many places where one feels close to nature. But it is the nature of a 'garden', in the widest sense, that the British are encountering.

For many centuries there have been places where human exploitation has not just changed nature but ravaged it. The Romans created much of the North African desert lands by their exploitive overcultivation. What is different today from the experience of past centuries is that human intervention is now on such a scale that it can affect the whole global environment, not just patches of it. Global warming can raise the sea levels around the world (though that will certainly not produce changes different in kind from those induced in the past by the natural thermal ebb and flow of the ice ages. Twenty thousand years ago, sea-level was 120 metres below its present height!). The destruction of parts of the ozone layer by fluorocarbons, with the consequent human exposure to lethal ultraviolet radiation, affects large tracts of the inhabited Southern Hemisphere. The slow accumulation of undegradable pesticides such as DDT, produced the silent spring as bird populations plummeted. These large-scale changes can creep up on us unobserved, and when detected they require significant periods of time for their reversal, even when resolute corrective action is actually being undertaken.

Population growth

All such problems, whether of the generation of hazard or of the control and correction of undesirable environmental impact, are greatly intensified by the continuing growth of the world's population. We have run out of room on spaceship Earth. There are no longer virgin lands to which to send surplus population. Of all the difficulties in the contemporary human encounter with nature, it is, I think, the continuing population explosion that is the greatest and the gravest. The pressure of people produces much pollution, and many environmentally destructive practices are pursued for short-term gain in order to cope with the increase in numbers of people.

In many international environmental conferences, whether political or ethical or religious, there is a reluctance to acknowledge that this is the case. Politically the difficulty arises from the fact that it is the developing, rather than the developed, nations that generate most of the population increase. People who are poor are bound to see children as one of the few resources to which they have ready access. The uncertainties of infant mortality in some less developed countries encourage the large family as an insurance policy, so that there is some surviving provision for meeting need in old age. Modern contraceptive practice is not always readily compatible with indigenous cultures. The developed countries cannot merely tell the developing countries to stop producing children It seems clear that a degree of prosperity is necessary before family planning becomes a widely acceptable practice. The only political way to implement a control on the Earth's population is to produce a fairer sharing of its resources.

Ethically the difficulty arises from the profound intimacy and significance associated with all individual human choices in relation to the family. The procreation and the care and nurture of children is one of the most deep and fulfilling activities of women and men. How can the community, whether the state or the whole world, take it upon itself to intervene in these personal decisions? All ethical thought encounters a tension between individual freedom and the common good, between the rights of persons and the needs of the society in which they live. This tension is particularly acute in relation to problems of population.

Religously the difficulty arises from differing theological interpretations of the nature of sexuality and the permissibility of different kinds of birth control. In particular, the Papal teaching in *Humanae Vitae* that artificial contraception is unacceptable poses problems, not only for the Roman Catholic communion, but also for other Christian communions that feel a certain delicacy in what they say in relation to the official position of the largest Christian church. The situation is further complicated by the obvious fact that very many Roman Catholic couples are not disposed to accept official teaching on the matter and that very many Catholic priests are clearly not pressing the point.

All these difficulties inhibit effective response to the problem of population growth, but nothing alters the fact that the containment of that growth is an absolute necessity, either in a planned or in a painful way, as part of the sustainable encounter of humanity with nature.

Gaia

There are some who say that we should not worry over much. They sometimes appeal to the Gaia hypothesis. This originated in the thought of James Lovelock,[4] who pointed out that the Earth has many homeostatic systems in operation, which have maintained an astonishing degree of balance in the circumstances necessary for the continuation of life, over periods of hundreds of millions of years, despite many fluctuations of terrestrial circumstances. For example, the oxygen content of the atmosphere, and the mean temperature, and the salinity of the seas have all been kept within tolerable limits. Many of these homeostatic mechanisms are not well understood, but their effectiveness is apparent. Perhaps a trifle incautiously, Lovelock made the comparison of the Earth and its biosphere to the integrated and regulated life of an organism, to which he gave the name of the goddess Gaia. In fact the Earth looks neither like a machine nor like a living being, but like some form of self-regulating entity lying in between, for which we have no appropriate model name to offer. It would be unwise to rely on the past as a guide to the future in supposing that all our environmental problems will find some natural solution. Remember that human nature has been operating for only a brief period, and on a globally significant scale for an even briefer period. Those who want to press the Gaia hypothesis to the limit may well wish to ponder whether humanity is not acting like some destructive virus within the Gaian body and that her immune system may well dispose of us by allowing human self-destruction!

Predictive predicament

The wealth of partly recognized, only partly understood, regulatory mechanisms to which Lovelock drew our attention, illustrates how complex is the living system of the Earth. In consequence, model calculations purporting to prophesy prospects for global change and to evaluate the results of such changes are notoriously unreliable. Too many feed back loops are involved for us to be able to estimate with any confidence. If global warming raises the temperature of the seas, more water will evaporate from their surfaces. This will produce greater cloud cover, which will blanket out some of the heat received from the sun. What is the net balance of thermal effect resulting from these linked processes? Even estimations of much simpler problems, such as the length of time left during which we can enjoy the resource of some particular fossil fuel, have proved subject to considerable error. In 1908, Theodore Roosevelt was told by one of his advisers that the United States of America would run out of anthracite within thirty years and of timber within fifty. The first members of the European Economic Community, often referred to as Club of Rome, with their dismal forecasts in the early 1970s, were also found to have called 'Wolf!' prematurely. The discovery of further reserves through geological exploration, and unforeseen developments in the pattern of use of fuels, have all drastically revised the basis on which the calculations had been made. Of course, fossil fuels cannot last for ever and I am not counselling complacency. Optimistic projections are likely to be as misleading and ill-founded as pessimistic ones. I am simply drawing attention to the difficulties of assessing what is going on.

Ethical debate

I think that warning is particularly necessary because of the character of much of the debate about the care of the environment that is taking place today. It seems to me that much of its tone is unduly shrill. People claim concerning some new development – whether it is nuclear power, or genetic engineering, or some change in agricultural practice – either that it is the best thing ever (with sliced bread often being proffered as the curious comparator) or the worse thing ever (so that immediate disaster looms).

Such a polarized debate cannot be helpful in our attempts to find a way to care for creation. Almost all the problems appear to be too complex to permit this kind of simplistic analysis. Almost always there is an account of potential gain and potential loss that needs to be reckoned up with the greatest care and scrupulosity. Yet our society fails to encourage rational debate of a balanced kind. The media thrive on the artificial construction of confrontation. If nuclear power is on the agenda, television will pit a representative of the nuclear waste disposal giant BNFL against a representative of Greenpeace, rather than organizing a round table discussion of the issues. The moral philosopher Alasdair MacIntyre[5] has drawn our attention to the fact that the absence in contemporary society of an agreed moral basis for decisions has reduced ethical debate to the strident assertion of individual opinions. If there is no common ground on which the disputants can meet, then the only strategy left is to try to shout louder than the rest. Hence the characteristic form of contemporary ethical debate is through the activities of single-issue moral pressure groups. I do not think this is healthy or conducive to the wise decisions and costly actions that are necessary if we are truly to care for nature. I recognize that there will be many in the Western world who will not be able to follow me in seeing the Earth as creation as the moral basis for a concern for the environment, but I think it is essential that Christians and other religious people should seek what common ground they can with all other people of good will in trying to articulate an ethical basis for caring for our world. Perhaps that common ground can be found in the acknowledgement of a respect for all humanity and for life and for the world that gave us birth. We need a shared concept of the common good, wide enough to embrace the natural world and future generations.

Science's contribution

In an informed and responsible debate, science and the scientists have an indispensable role to play. To many that will seem like recruiting Bill Sykes into the police force or getting an arsonist to join the fire brigade. Second only to Christianity, and before it in many people's minds, science is seen as the villain of the environmental piece. Is it not the unconstrained use of technological power, made available through scientific discovery, which has enabled human intervention in nature on a globally disastrous scale? The suspicion of experts and their advice has led the general public to reject sensible programmes to deal with some difficulties. It seems likely that the decision not to sink the Brent Spar oil platform in the deep Atlantic was an injudicious decision of this kind.

The true story is more complex and less gloomy than these criticisms suggest. Of course, there *are* dangers. The beginning of wisdom in considering the ethical use of scientific discoveries is to recognize their ambiguity. The power they provide can be exercised either for good or for ill. We may rightly fear the eugenic use of genetic manipulation but rightly rejoice at the prospect of the same knowledge being used to bring about a cure for sufferers from monogenetic diseases such as cystic fibrosis. The recognition of this double-edged status of scientific advance does not mean that questions of right use are not the concern of the scientists so that they can just be left to society to decide. Science is not value free in the sense of being ethically neutral with respect to questions of the exploitation of its discoveries. That is because it is only the experts who can attempt some informed estimate of the possibilities and consequences resulting from the use of scientific understanding. But equally, the decisions cannot be left to the experts alone, for their judgements of moral rightness are not peculiarly privileged. Their insight is as open as that of anyone's to ethical distortion and so the decisions need to be taken in a wider setting than that provided by the research team.

There is a Charles Wesley hymn that asks that we should be delivered from 'our calling's snare'. All callings have their snares. For the scientist, the temptation is to give way to what one might call 'the technological imperative'. In the excitement of discovery it is only too easy to get caught up and carried away with the desire to take the next step and to see what happens, without stopping to ask whether one should really go ahead. Not all that can be done should be done. It is important to ask the ethical questions at an early stage and not after the technology has arrived on the shelf ready for use.

If you read the reminiscences of those who were at Los Alamos during the war, you will find that for nearly all of them, they only asked themselves what they were doing when they saw the first test explosion in the New Mexico desert. On the top of that mesa, there had been collected perhaps the greatest concentration of scientific talent ever brought together for a common purpose. As the physicist Robert Oppenheimer said 'the science was sweet' and they just got on with it. I am not saying that they should not have made the atomic bomb (in fact, I think it was a justified project) but they should have reflected earlier on what they were doing. Yet scientists are not by nature heedless and irresponsible. The 'mad scientist' of popular imagination and story, intent on discovery at any cost, is fortunately extremely rare. The history of experimentation with recombinant DNA is one of a community aware of potential dangers and anxious to avoid them, accepting a careful (and initially voluntary) regulatory control over their activities.

If we are to make wise decisions we need continuous interchange between the scientists and wider society. The public has to say 'Do you really know what you are doing? Have you thought about its wider and longer-term implications?'. The scientists have to be careful and scrupulous in evaluating the potential gains and losses that can result from a particular line of work. Only the scientists have the technical resources to be able to try to produce an even-handed analysis of benefits and dangers. We greatly need arenas for informed discussion in which science and society can together seek careful and responsible policies.

Questions of economics cannot be absent from such a discussion, but the cost–benefit analyses presented must take all effects into account. I am told that acid rain causes the loss of timber to the value of £16 billion per year in Europe, and this certainly needs to be taken into account in assessing the costs of its prevention. If Californian brambles are allowed to grow and provide homes for a certain kind of wasp, that useful creature will, by preying on pests which affect the grape vines, save the farmers $US125 per acre per year in expenditure on pesticides.[6]

There is also the possibility that scientific advance will afford new and successful answers to old and difficult questions. Many problems, both in relation to the use of resources and also in relation to impact on the environment, arise from growing human needs for energy. Whatever measures of conservation and restraint may be undertaken in developed countries, it is clear that the world's energy requirements overall will continue to grow as the developing countries gain access to the resources that they need. If science and technology could solve the problem of nuclear fusion, an abundant uncontaminated energy source would be created, free from radioactive by-products and using water as its basic raw material. Nuclear fusion has provided a particularly intransigent problem to solve. During the whole of my scientific life, the optimists have been predicting that just another twenty years will bring the necessary breakthrough. Such a moving horizon of hope has proved a little discouraging, but surely some day we will find out how to tame the energy of the hydrogen bomb and make it available in useful forms. That discovery, when it comes, will profoundly affect human life and modify in beneficial ways its impact upon the natural environment.

Difficult as right judgements are to attain, it is even more difficult to secure their implementation. A concerned debate within society must result in the political will to attain desirable ends. Whatever the internal tensions, justice, peace and the integrity of creation do belong together. In a free society, many of the most effective political levers are fiscal in character. The principles that the polluter pays and that conservation can be achieved through taxation are among the most effective that we have at our disposal, if we have the will to use them.

The end of the matter

It is difficult to bring a chapter of this kind to a resounding conclusion. I have tried to survey an intrinsically complex situation. The caring and responsible interaction of humanity with created nature is vital to achieve but its attainment is difficult and there are many perplexities about what strategies should be pursued. Knowledge is an indispensable basis for environmental thought. We have to seek as full and unbiased an assessment of risks and opportunities as we can possibly attain. Science must be an active participant in the dialogue, for otherwise we shall be groping in ignorant darkness. Change is inevitable, but as far as we are able, its consequences must be assessed and the results continuously monitored. But decisions in these matters depend upon the perception of value as well as the recognition of fact. I believe that humanity divorced from nature is always in danger of becoming brutalized and self-destructive. Human beings need to evince what Albert Schweitzer called 'a reverence for life'. Among the ranks of humanity, scientists, with their sense of wonder induced by their encounter with the order of the physical world, and with the knowledge that that encounter has given them, have a significant role in helping society to behave responsibly, in terms of both present practice and of provision for future generations.

NOTES

1. L. White, 'The historical roots of our ecological crisis' *Science*, 155, p. 1203.
2. P. Hefner, 'The Evolution of the Created Co-Creator' in T. Peters (ed.), *Cosmos as Creation*, pp. 211–33 (Abingdon Press, 1989).
3. A. Linzey, *Christianity and the Rights of Animals* (SPCK, 1989).
4. J. Lovelock, *Gaia* (Oxford University Press, 1979).
5. A. MacIntyre, *After Virtue* (Duckworth, 1981).
6. I learnt these examples from a lecture by Professor Sir Ghillean Prance.

Index

Printed in the United Kingdom
by Lightning Source UK Ltd.
2816